Support Technology for Transportation Environment of Handicapped People

Support Technology for Transportation Environment of Handicapped People

Hiroshi Ikeda

Union Press

First published in 2013
by Union Press®
(a division of Union Services Co., Ltd.)
1-6, Uehommachi-Nishi 5 chome, Chuo-ku, Osaka, 542-0062 Japan
www.union-services.com

DTP by Design International
Printed and bound in Japan by Shinano Co., Ltd.

ISBN 978-4-946428-64-7

Printed in Japan

Contents

List of figures x
List of tables xiv

Introduction 1

CHAPTER ONE Supports for persons with disabilities 5

- Services and supports to improve the self-reliance of persons with disabilities 5
- Changes in the understanding of self-reliance 6
- Normalisation 6
- Persons with physical disabilities as welfare targets, and services and support for them 7
- Hierarchy of self-actualisation 8
- The value of social welfare 9
- Nursing care service 10
- Nursing care professionals and their workload 11
- Heart Building Law in Japan 13
- Barrier-free Transportation Act in Japan 13
- New Barrier-free Act in Japan 14
- Guidelines for the maintenance and improvement of easier accessibility and usability of transportation infrastructure 15

Historical background of universal design 16
Services and Supports for Persons with Disabilities Act 17
ADL and IADL 18
Quality of Life 19
Rehabilitation 20
Psychological impact on persons with congenital disabilities 22
Acceptance of disability by people who acquired disabilities and developmental disabilities 22

CHAPTER TWO Physical and mental features of persons with a limb/trunk dysfunction 25

Definition of the term "disability" in Japan 25
Physical features of persons with limb/trunk dysfunctions 26
Psychological features of persons with limb/trunk dysfunctions 28
Spinal-cord injury 28
Progressive disorders 32
Cerebral paralysis 33
Amputees and artificial parts of the body 35
Disuse syndrome 36

CHAPTER THREE Wheelchairs and barrier-free sidewalks 39

The history of the development of wheelchairs 39
Design and product configuration 43
Models and configurations of wheelchairs 44
Electric wheelchairs and Japanese Industrial Standards 54
Sidewalks 57
Footsteps and slopes 64

CHAPTER FOUR Wheelchair users and cars 67

The environment for driving cars 67
Acquisition of a driver's license by persons with physical disabilities 69
Special marks for persons with physical disabilities 71

Parking spots for the disabled and the system to exempt the disabled from parking prohibition rules 72
Subsidy programs 74
Assist devices for driving which are tailored to each type of impairment and the selection of vehicles 75
Hand-operated assist-devices for driving 75
Driving seats 79
Turning devices 81
Foot-operated assist-devices for driving 82
Transferring 84
Support tools 85
Problems in driving vehicles experienced by persons with physical disabilities 87
Remodelling and financial aid 95

CHAPTER FIVE Physical features of persons with a hearing disability 97

The time when a hearing impairment appears 97
Types of hearing disability 98
Differences in the nature of hearing loss 98
Senile hearing loss 100
Classification of impairments in hearing ability 101

CHAPTER SIX Communication and transportation of persons with a hearing disability 103

Hearing aids and artificial ears 103
Types of hearing aids 104
Digital hearing aid 107
Wearing hearing aids on both ears 109
Purchase of a hearing aid 110
History of the hearing aid 111
Methods of communication 113
Information obtained visually 115

Disadvantages in everyday life 116
The environment for driving cars 117
History of the campaigns for the acquisition of a driver's license 118
A special mark for persons with hearing disabilities 120
A license system in relation to acoustic sense 121
Difficult traffic conditions for persons with a hearing disability 122
Information gathering by using acoustic and visual senses during driving 123
Use of a hearing aid during driving 123

CHAPTER SEVEN Physical features of persons with visual impairment 125

Types of visual impairments 125
Ageing-related visual disabilities 128
Classifications of visual impairment, and the walking of persons with visual impairment 130
The timing of occurrence of visual impairments 131

CHAPTER EIGHT Transport of persons with visual impairment and their daily lives 133

Walking with a guide 133
Walking alone by using a white cane 134
Walking with a guide dog 135
Burdens received by visually-impaired people while walking alone 136
Symbol marks for persons with visual impairment 137
Braille block 139
Valid width of sidewalks 141
The difference in levels on pedestrian crossings 142
White canes 143
Paying attention to people walking with a white cane 144
History of guide dogs 145
Traffic-light machine with an acoustic device 146

CHAPTER NINE Public transportation systems in Japan 149

Getting-on and getting-off trains, and the reserved space for persons with physical disabilities 149
Elevators and escalators at stations 151
Barrier-free buses 153
Demand-responsive buses and special transportation service 155
Light-rail transit 156
Park and ride system 158
Barrier-free airplanes 159
Subsidy for each public transportation system 159

Epilogue 161

References 165

Index 169

List of figures

The PDCA cycle advocated by William Edwards Deming 2
Figure 1-1 Theory of degrees of needs (desires) by Abraham H. Maslow 8
Figure 1-2 The importance of specialist jobs and regular jobs 10
Figure 2-1 Spinal cord nerve segments 29
Figure 2-2 Order-made artificial parts of a body 35
Figure 2-3 The regions where decubitus appears most commonly in the human body 36
Figure 3-1 A picture of a bed with wheels painted on a vase 39
Figure 3-2 A picture of a wheelchair drawn on a slate 40
Figure 3-3 The wheelchair used by Philip II 40
Figure 3-4 A wheelchair which was designed to be propel ed by hand 41
Figure 3-5 The Hakone-style wheelchair 42
Figure 3-6 Names for parts of a wheelchair 46
Figure 3-7 A wheel with an illustrated hub cap for children 48
Figure 3-8 Cushions on wheelchairs 49
Figure 3-9 A fixed-style armrest 50
Figure 3-10 Footrests, with one footrest pivoted upward 51
Figure 3-11 A lever brake-system 52
Figure 3-12 A grip with a brake 52
Figure 3-13 A tipping lever 53
Figure 3-14 An electric wheelchair with a joystick 55
Figure 3-15 A mobility scooter 56
Figure 3-16 JIS mark with additional welfare tool mark 56
Figure 3-17 TS mark 57
Figure 3-18 Rolling of sidewalks 58
Figure 3-19 A stair lift 61
Figure 3-20 A puddle of water made in the depression on a sidewalk 62
Figure 3-21 The concave-convex surface of interlocking blccks 62

Figure 3-22 Clogged grating holes with fallen leaves ... 63
Figure 3-23 Wheelchair climbers ... 64
Figure 3-24 Flat area on a ramp ... 65
Figure 3-25 The difference in distance of movement of parts of the body in various gradient ratios on ramps ... 66
Figure 4-1 Process of obtaining a driver's license for the disabled in Japan ... 70
Figure 4-2 Mark for persons with physical disabilities ... 71
Figure 4-3 The International Symbol of Access for persons with disabilities ... 72
Figure 4-4 A parking spot for the disabled ... 73
Figure 4-5 Card which indicates that the vehicle is exempted from parking prohibition rules ... 73
Figure 4-6 A device with a hand-control system known as a floor type ... 77
Figure 4-7 A hand-operated driving-device known as a column type ... 78
Figure 4-8 Hand-operated assist-devices for driving which have been developed by auto-makers ... 78
Figure 4-9 A joystick type assist-device for driving ... 79
Figure 4-10 A driving seat for persons with physical disabilities ... 80
Figure 4-11 A body belt with a loop fastener ... 80
Figure 4-12 Turning devices ... 81
Figure 4-13 A foot-operated device ... 82
Figure 4-14 An accelerator-pedal for left foot ... 82
Figure 4-15 Combination switches for the foot ... 83
Figure 4-16 A device which helps the driver fasten the seat-belt automatically ... 83
Figure 4-17 A transfer-board ... 84
Figure 4-18 A wheelchair lift ... 85
Figure 4-19 An electric wheelchair lift ... 85
Figure 4-20 A lift-up passenger-seat ... 86
Figure 4-21 Transportation vehicles equipped with assist devices ... 86
Figure 4-22 Problems caused by centrifugal force ... 88
Figure 4-23 Body parts which drivers with impairments wish to have fixed in place ... 89
Figure 4-24 The movements of a driver's head at different speeds ... 89

Figure 4-25 The impact of centrifugal force when driving on curves 91
Figure 4-26 The amount of muscle activity in different parts of the body when driving on curves 92
Figure 4-27 Driving posture when driving on curves 94
Figure 4-28 The amount of money paid by individuals to remodel cars 96
Figure 5-1 The anatomy of the ear 99
Figure 6-1 A hearing aid for inserting in the ear 105
Figure 6-2 A hearing aid for hanging on the ear 106
Figure 6-3 A hearing aid designed to be put in a pocket 106
Figure 6-4 A spectacles-type of hearing aid 107
Figure 6-5 The degree of hearing loss and suitable hearing aids 108
Figure 6-6 The actions of the right and left brain 110
Figure 6-7 The mark indicating that a shop is certified by ATA 111
Figure 6-8 An ear-trumpet 111
Figure 6-9 An advertisement for the hearing aid designed by Koukan Shiba 112
Figure 6-10 The mark for drivers with hearing disabilities 120
Figure 6-11 The mark called an "ear mark" in Japan 121
Figure 7-1 Characteristics of how patients with eye diseases see 127
Figure 7-2 Causes of glaucoma 130
Figure 8-1 The basic way of leading people with visual impairment to necessary places and activities 134
Figure 8-2 Mean relative heart rates 137
Figure 8-3 The universal symbol for visually-impaired people 138
Figure 8-4 A guide-dog mark 138
Figure 8-5 Braille block 139
Figure 8-6 Types of Braille block 140
Figure 8-7 Manhole cover and Braille blocks 141
Figure 8-8 Illegally parked bicycles and shop displays near Braille block on a sidewalk 141
Figure 8-9 The difference in level on pedestrian crossings 142
Figure 8-10 A guide dog 145
Figure 8-11 The push-button of a traffic-light machine with acoustic device 146

Figure 9-1 A wide ticket-gate 149
Figure 9-2 A call-button on a platform 150
Figure 9-3 The reserved space for wheelchairs in a train 151
Figure 9-4 The mirror and special control board inside an elevator 152
Figure 9-5 A call-button placed at the entry of escalators 153
Figure 9-6 The entry of a non-step bus 154
Figure 9-7 An old type of tramway in Japan 157
Figure 9-8 Unmanned car par 158

List of tables

Table 2-1 Spinal-cord injury levels and ADL 30
Table 3-1 Classification of wheelchairs 45
Table 3-2 The dangerous places and the types of injuries 54
Table 4-1 Assist devices for driving tailored to each type of impairment 76
Table 5-1 Classification by decibels of persons with hearing disability 101
Table 5-2 Grade of disability for persons with a hearing disability 102
Table 6-1 History of the campaigns for the acquisition of a driver's license for persons with a hearing disability 119
Table 6-2 The types of vehicles which can be driven by persons with a hearing disability 122
Table 7-1 Different classes of persons with visual impairment 125

Introduction

Today, as a social issue, there has been a lot of focus on making roads and facilities barrier-free in order to support the independent life of persons with physical disabilities. According to the 2012 annual report on government measures for persons with disabilities, the number of persons in Japan with disabilities amounts to 7.443 million. The number in the report includes persons with physical disabilities (including visual impairment, hearing disability, speech disability, limb/trunk dysfunction and internal impediments), persons with intellectual disabilities and persons with mental disabilities, and their numbers are 3.663 million, 0.547 million and 3.233 million, respectively [(1)]. In addition, the number for the elderly—comprising people at 65 years-old and over—is 29.75 million, which shows the rate of ageing amongst the Japanese population has increased by 23.3 %—according to the 2012 white paper on the aging of society [(2)]. Along with this, we can see an increase in the number of people suffering difficulties with walking caused by lifestyle-related diseases caused by ageing. The greying of society is expected to continue, and if the number of persons with walking difficulties is added to the total population of persons with disabilities in general, then this figure certainly becomes a large proportion of the entire population.

Against this background, some social infrastructure related to the creation of a barrier-free environment has so far been implemented. Until recently, that infrastructure was said to be more than 30 years behind in comparison to other

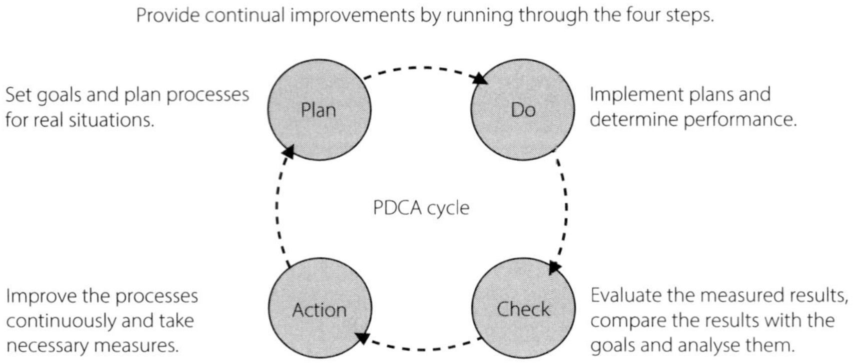

The PDCA cycle advocated by William Edwards Deming

advanced countries. However, people have come to realize the importance of maintaining and improving a "gentle living environment for people" corresponding to the physiological and psychological needs of persons with disabilities. Since the Barrier-free Transportation Act came into force in November 2000, a barrier-free transportation environment has been created and the means have been developed to fit this to persons with disabilities. Furthermore, a concept of everlasting improvement has been implemented by using the method known as the PDCA cycle, and a system to evaluate what was developed has been introduced. The PDCA cycle is a method which provides continual improvements to tasks by running through four steps: Plan → Do → Check → Act. The figure above shows the details of the cycle.

Though Japan's maintenance of, and improvements to, its living environment have not yet caught up with Europe, the most advanced area in the developed world, the gap between Japan and Europe has rapidly narrowed. For instance, Japanese guidelines relating to such maintenance and improvements now compare favourably with the standards of European countries.

However, carrying out maintenance and improvements only on the hardware side will never be a fundamental solution to the overall problem. A balance between the hardware and software sides is a key for any kind of function to work well. In the case of Japan, "removing barriers in people's mentality" is important as a development on the software side. For instance, when Japanese people see persons with a visual impairment in difficulty on the street, many find themselves not knowing how to offer support. One of the reasons why this happens is because Japanese people generally have not had much social education in learning how to deal with such a situation. By contrast, when people in other advanced countries are involved in activities in their communities, they also engage in volunteer work. This helps people's spirit of volunteerism to be cultivated. In many cases in Japan, people learn about volunteerism only as a school subject. Therefore, people need to learn about volunteerism at schools, societies, homes and any other opportunities in the future.

For example, if you see a person in a wheelchair who is in trouble because of steps in front of her/him, one of the ways to support her/him is to offer help from behind. A 5 cm difference in levels at the entrance of a building, for instance, is not of any concern to healthy people, but for people in a wheelchair it is a big barrier which prevents them from moving ahead. If you understand the nature of such difficulties for persons with disabilities, you may be able to support them spontaneously. When you see that persons with disabilities are facing problems or difficulties, what is important for you is to "help" them: to begin with by, for example, saying something to them. Japan may change greatly if people can remove barriers from their mentality, create a climate in which it is a natural thing for them to help persons with disabilities and create social systems which support such a spirit among people. In addition, it is important for healthy people to get involved more willingly with persons with disabilities and elderly people.

This book summarises the current situation of the transportation environments for persons with limb/trunk dysfunction, hearing disability and visual impairment, while also describing the problems which nowadays exist.

In chapter one, the Japanese law and support system for handicapped persons is introduced. Especially, it focuses on support for independence. In chapter two, the physical and mental features of persons with a limb/trunk dysfunction are summarized. Chapters three and four summarize the transportation environment for persons with physical disabilities. In chapter three, the history and structure of wheelchairs and support equipment to be used for wheelchair users, and also, the transportation environment of sidewalks, footsteps and slopes are discussed. In chapter four, assist devises and the acquisition of a driver's license for persons with physical disabilities are introduced. In addition, the difficulties experienced at the time of driving are discussed. Chapters five and six summarize the physical features, hearing aids and commutation methods for persons with a hearing disability. Chapter six also discusses the difficulties for persons with a hearing disability when they drive a car. In chapters seven and eight, the physical features of persons with visual impairment are summarized. Chapter eight also introduces Braille blocks and guide dogs, and discusses of walking problems on the sidewalk. Finally, chapter nine summarizes public transportation systems of trains and buses in Japan as transportation methods for handicapped persons.

CHAPTER ONE

Supports for persons with disabilities

Services and supports to improve the self-reliance of persons with disabilities

In present-day Japan, a new range of welfare programs has been implemented in addition to the traditional benefits-style welfare programs of the past. This coincides with global trends. The new programs are in accordance with the concept of "community welfare", and they aim to provide support to persons with disabilities and elderly people through activities conducted by "the wider world, towns and societies" while assuring their social revitalisation.

More specifically, the range of problems relating to the usage of roads by persons with disabilities and elderly people is one of the high-profile issues in Japan. The mobility of the disabled and the elderly needs to be secured, since mobility is essential for them to be independent. Similarly, the structures known as "the welfare-support information and communication system", "the attendant service and at-home-care support system", "the social-welfare image-information delivery system" and various other projects have all been established in the information and communication fields as government initiatives to prepare the environment so that persons with disabilities and elderly people can have independent lives. Moreover, a "long-term care insurance system" has been launched together with other welfare policies in an effort to improve the self-reliance of persons with disabilities and elderly people.

Changes in the understanding of self-reliance

Our understanding of the concept of the self-reliance of persons with disabilities has improved, and significant changes have been made since the war, coinciding with changes in the social economy. During the period of economic reconstruction, that is, shortly after the war through into the1960s, a period when a priority was placed on economics, if people discussed the self-reliance of persons with disabilities, it meant only their economic and physical independence. Accordingly, persons with severe disabilities who had difficulties with daily activities were considered as persons subject to protection.

After the 1970s, the Independent Living campaign spread in America, while the concept of "normalisation" began and spread from Denmark. Since then, much debate has been carried out with a view to considering all persons with severe or less than severe disabilities as subjects for an independent life, and more persons with severe disabilities have aimed to have an independent life in their community. Meanwhile, there has been an increase in the number of groups which campaign for promoting self-reliance and social participation by persons with disabilities, and the principle of normalisation spread rapidly following the International Year of Disabled Persons in 1981.

Japan has changed from a society that places its first priority on economics to a society that places an importance on people's lives. With this change, the time has come for us to reconsider our understanding of the self-reliance of persons with disabilities and to change our understanding of their self-reliance from the view that independence means mere economic and physical independence to the view that independence for them is their self-actualisation while receiving health, medical and welfare services in their communities.

Normalisation

Normalisation is a social principle related to the welfare of society, and the principle started in North European countries in 1960. According to this principle, a normal society is a place where no distinction is made between persons with disabilities or elderly people and healthy people, and it is also a place where they live together while helping each other. This principle also includes campaigns and programs of which the purpose is to realise such a society. In the 1950s, a group of parents of children with intellectual disabilities discovered that many violations of human rights were being committed in a giant facility for disabled people in Denmark. As a result, the parents launched campaigns so that they could improve such terrible conditions. The principle of normalisation took root in this incident.

The advocate of the principle, Bank Mikkelsen, described normalisation as follows: it is to offer persons with disabilities the same living circumstances as normal people; it is not to try to make the circumstances of persons with disabilities close to the circumstances of the normal, but rather to make circumstances in which the normal conduct daily routines usual also for persons with disabilities; and, it is for normal people to act in a way that they would want other people to do for them if, or when, they themselves became persons with disabilities (3).

In 1993, the Japanese Basic Act for Disabled Persons was implemented based on the principle of normalisation. However, Japanese society has not been very successful at embodying the principle. The principle of normalisation needs to penetrate and spread further in order to bridge the gap between the principle and reality in society.

Persons with physical disabilities as welfare targets, and services and support for them

When we think about persons with physical disabilities, we tend to pay attention only to their impairments. However, it is very difficult for people with disabilities to recover their physical functions completely. Therefore, if attention is drawn only to their impairments, it will not solve their problems. The final goal of welfare services for persons with disabilities is their "self-reliance." The aim of the welfare service for persons with physical disabilities is not to solve their condition of disablement, but to help in solving the difficulties which they face in their daily routines.

Supporting the self-reliance of the disabled is the role of aid providers, and the following five bases are vital in providing such services and support.

(1) Services should offer enough variety for the users to choose.
(2) Services should be selected by the users themselves.
(3) The users' place of residence has to be secured within their community and society, unless it is impossible.
(4) Independent mobility and means of communications for the users should be secured, and the users should be able to participate independently in social, economic, cultural and a variety of other activities while maintaining good inter-human relations in their community.
(5) The users should be able to participate in a variety of policy-making processes.

Hierarchy of self-actualisation

The mental burden on persons with disabilities is enormous. Thus, it is very important that they accept their disabilities, while they are being provided with support, so that they can take a positive attitude towards their life. Abraham H. Maslow [(4)] has analysed the hierarchy of needs (desires) of human beings, and they are shown in Figure 1-1.

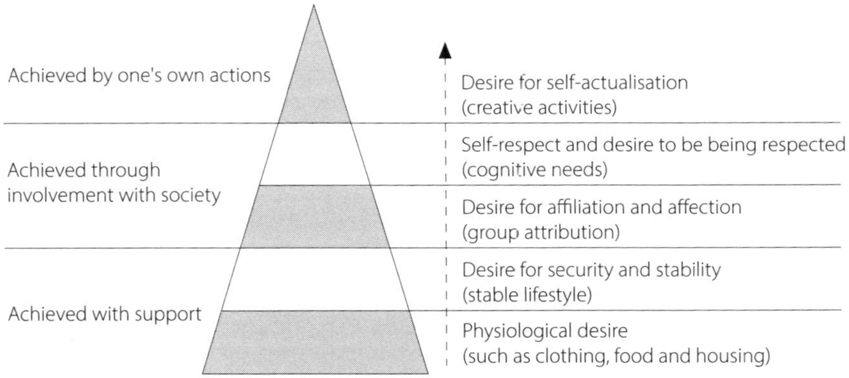

Figure 1-1 Theory of degrees of needs (desires) by Abraham H. Maslow

At the bottom of the hierarchy are "physiological needs" and the "desire for security and stability", which are the fundamental needs (desires) for survival, and these needs can be satisfied by support from other people. However, "self-respect and desire for respect" and "desire for affiliation and affection" can be realised only by having relations with society, and these desires can be satisfied only when people are willing to live positively. Satisfaction with such desires will lead to having a "desire for self-actualisation", and eventually people will start taking action by themselves. Once physiological desire is satisfied, people become interested in fulfilling social desires such as group attribution, affection and approval. Once these desires are satisfied, the desire for self-actualisation can be fulfilled.

Self-actualisation is not related to physical self-reliance but the key point is to have clarity in one's own thinking and to be mentally independent. Therefore, when persons with disabilities receive assistance from caregivers and supporters, these persons should be able to decide what kinds of assistance they wish to receive, and their decisions ought to be respected by their caregivers and supporters. Self-reliance includes:

(1) Economic independence
People are capable of maintaining the minimum-standard of living.
(2) Independence in one's own daily affairs
People are capable of taking care of their own daily affairs.
(3) Mental independence
People are capable of living according to their own thinking.
(4) Personal independence
People are capable of living according to their own way of thinking while receiving support from other people.

The value of social welfare

Occupations related to such fields as medical care, law, education and religion are thought of as specialist jobs, and they are usually different from regular jobs. People who engage in social welfare are considered as working in this area of

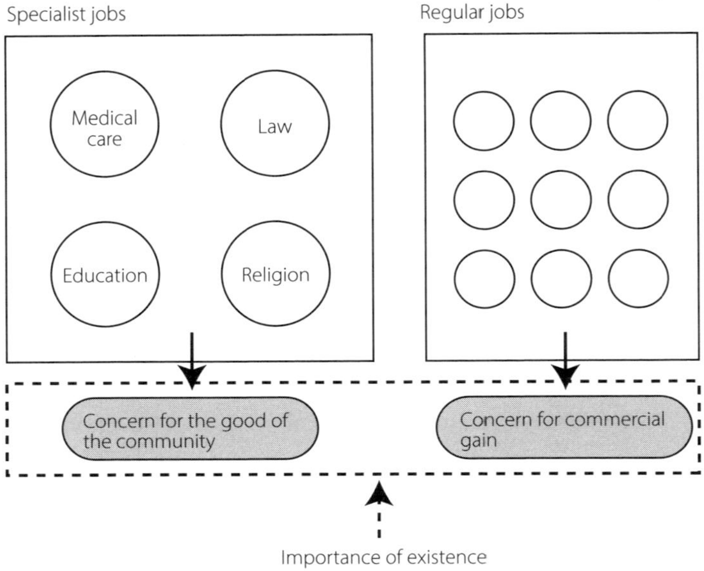

Figure 1-2 The importance of specialist jobs and regular jobs

specialist jobs. As is shown in Figure 1-2, the fundamental difference between such specialist work and regular work is whether the nature of its goal is communality or commercial gain. Regular jobs may also require highly-advanced knowledge or skills, but after all, their final goal is the pursuit of interest. However, according to Akito Kurokawa [(5)], the importance of work in medical care, law, education and religion in society is to offer special knowledge and skills to society regardless of one's interests.

Nursing care service

Care for persons with disabilities and elderly people has been traditionally provided by families. The common sorts of care provided by family were nursing care and livelihood protection, and in Japan the burden on the family members was traditionally shared by three generations: grandparents, parents and children. However, because the family structure has changed due to the declining birth-rate, the ageing population and the trend toward nuclear families, the families which provide care have limits placed on their efforts to bear the burden which rests upon them. This is one of the social issues we face today in Japan. Against this background, promotion of the training of people as welfare caregivers and as other types of specialists has been launched following the implementation of the Ten-year Strategy for Promoting Geriatric Health and Welfare, the so-called Gold Plan in 1989. When people—other than immediate family members—provide care for elderly people and/or persons with disabilities, the caregivers assume a social responsibility. Thus, they are required to have knowledge and skills that measure up to certain criteria, and also they are required to be properly certified.

The term "nursing care" was originally defined in an ordinance when a law relating to pensions given to disabled veterans of the army was implemented in Japan in 1892. The concept of nursing care existed first as a criterion for benefit payments of pensions. However, several campaigns were carried out by persons with disabilities in a demand for guaranteed nursing care in the latter half of the 1970s, and since then the use of the term "nursing care" has become more general.

Nursing care professionals and their workload

In Japan, nursing care has been provided by the female members of each household and traditionally it is a form of unpaid work. In April, 2000, the Long-term Care Insurance Act was implemented. Since then, the number of professional care workers has increased rapidly. However, the Long-term Care Insurance Act was amended in 2006, and it led to a new approach termed "care prevention" by the government in its ordinance. The purpose of this new approach was to prevent many elderly people from needing and/or receiving care under the long-term care systems which were stipulated by the Long-term Care Insurance Act. Accordingly, application for the home-care service was confined to persons with severe disabilities. As a result, the number of home-care workers has actually decreased. The reasons for the decrease are: it is one of the hardest jobs in nursing-care-related work and its wages are very low; and the owners of home-care-related businesses have cut down on the number of their workers due to the decline of profit margins caused by the undercutting of nursing care benefits [6]. Thus, although one of the difficult issues of replenishing manpower needed for supporting nursing care has been resolved by the introduction of Long-term Care Insurance, the working environment for caregivers has further deteriorated because of the undercutting of nursing-care benefits. It is still difficult to create a stable employment environment under these circumstances.

Nursing care involves exhausting labour, such as lifting care recipients, and the workers tend to suffer from mental stress. In nursing-care facilities, the spirit of dedication often takes priority over the workers' physical and mental health. Unsurprisingly, a large number of caregivers suffer from backaches, and it often becomes chronic. This is one of the most serious problems among the workers [7]. The most common back problems suffered by caregivers are "muscular lumbago"—caused by chronic tension or fatigue—and "lumber disk hernia"—caused by heavy strain on the lumber disk. Both types of backache are related to slumping or other unnatural working postures as well as to lifting heavy objects.

Nursing care once drew much attention as an occupation. However, its harsh working circumstances have caused a decline in its popularity. The series of problems that occurred came about because of the government's desires, and its unfruitful efforts, to kill two birds with one stone, as they tried to resolve two problems of finance and the quality of helpers at the same time. This was done even though the priority had to be on securing a certain amount of service to be delivered. It is important to prepare adequate nursing care benefits and an adequate labour environment for caregivers so that they can work comfortably.

In Japan, there is a pervasive idea that everything has to be done by hand in the field of nursing care, because the care recipients are human beings. Moreover, some workers refuse to use electric lifts in the event of loading and unloading care recipients, and some claim that using lifts takes too much time and that it is easier to do the job by hand. Certainly, persons with disabilities—people who probably have less opportunity than usual to have direct contact with others—may feel more comfortable if they can sense the warmth of other people. However, some nursing care facilities are short of staff, and efficiency is very important in such places. Nursing care equipment is likely to increase the efficiency of work if it is used properly, and it will be effective if workers can judge precisely when to proceed by hand and when to use machines in their working environments.

Heart Building Law in Japan

The Heart Building Law was an act related to making buildings accessible to, and usable by, the elderly and physically disabled persons. In Japan, this act was implemented in September, 1994. The targets of this act were the elderly, persons with physical disabilities and any other people who had limits placed on their social activity because of some kind of functional impairment. The purposes of this law were: to take necessary measures in promotion of the creation of accessible and usable buildings; to improve the quality of specific buildings; and to improve public welfare.

The law aimed to set criteria for public buildings, such as railway stations, department stores and hotels, which many people, including persons with disabilities, use. This law was well received also by business enterprises and the number of companies which were accredited by the law increased from about 1995 onwards. However, as the New Barrier-free Act was implemented in 2006 (which will be described later in this chapter), the contents of the Heart Building Law were repealed in accordance with the Barrier-free Transportation Act. Thus, the Heart Building Law was effectively abolished.

Barrier-free Transportation Act in Japan

The Barrier-free Transportation Act—an act aiming for public transportation accessibility and usability by elderly and physically disabled persons—came into effect in November, 15, 2000. The drawing up of "fundamental plans for the facilitation of transportation" and the maintenance and improvement of walking spaces were actively advanced by local governments. In order to facilitate transportation accessibility and usability for all users, it was necessary to have effective as well as adequate plans and designs suitable for persons with disabilities or with difficulty in walking. Therefore, a sustainable amount of capital needed to be invested into the realisation of such plans. Purging barriers in transportation-related circumstances is very important not only for improving the safety of walking spaces for persons with difficulties, but also for expanding their area of daily activities and the opportunities they may have for their social participation.

Barriers on the roads, for example, increase the physical burden in walking unnecessarily for persons with difficulties. This negative circumstance confines the distance and frequency of walks they may take in the outside world from their homes, and also it confines their social activities in everyday life. When barrier-free environments are created, it is important to consider how to reduce physical burdens for users, including persons with disabilities and their caregivers, and how to ensure psychological comfort in addition to their safety.

New Barrier-free Act in Japan

The Barrier-free Transportation Act and Heart Building Law were integrated on December 20, 2006, and the New Barrier-free Act "for transportation accessibility and usability by the elderly and physically disabled persons" became effective. The barrier-free areas which were built based on the Barrier-free Transportation Act were like small dots scattered on a map, while the barrier-free areas built according to the New Barrier-free Act are more like a series of flat surfaces, which include outdoor parking spaces and city parks. In addition, the new act has extended its targets by including persons with intellectual, mental and developmental disabilities, and the related facilities have been improved according to the concept of universal design.

In order to create transportation environments with smoother accessibility and usability, the safety and user-friendliness of transfer pathways of facilities and of related vehicles need to be taken into consideration, and even facilities which do not directly have any obligations of creating easier accessibility and usability for persons with disabilities imposed on them are expected to take certain positive action in accordance with the act. Moreover, the delivery of easier, clearer and more appropriate information has been made compulsory in emergency situations. As for the educational training of staff or other related personnel in the public transport system, appropriate training is to be set up in order to prevent persons with disabilities from being rejected when boarding or using public transport systems and to ensure smooth delivery of services.

These fundamental strategies were instituted by the New Barrier-free Act and some councils were set up for implementation. In addition, two considerations were included within the meanings of creating easier accessibility and usability of transportation infrastructure within the priority areas of improvement. One consideration was harmony with the plans related to the improvement of welfare and urban performance set up by municipal governments. The other consideration was the usage of the council in order to reflect the views given by elderly and physically disabled persons regarding the plans. Furthermore, it was stipulated that the priority areas for improvement should have three or more facilities closely connected to the daily lives of persons with disabilities, and their transferral between or/among such facilities should usually be on foot. It was also stipulated that it is important to seek municipal cooperation when large-scale facilities exist in priority areas. Moreover, it was considered important to conduct adequate assessments of current circumstances, and to have suitable liaison and coordination in order to ensure smooth progress in the operations concerned. In addition, importance was placed on the participation of users, as well as citizens' evaluations conducted afterwards by the councils.

Guidelines for the maintenance and improvement of easier accessibility and usability of transportation infrastructure

Guidelines for the maintenance and improvement of easier accessibility and usability of transportation infrastructure were implemented in July, 2007 in relation to the passenger facilities of the public transportation system. These guidelines include details for the ideal maintenance and improvement of passenger facilities in order to achieve easier accessibility and usability of transportation infrastructure while at the same time meeting the needs of a diversity of service users, including elderly and physically disabled persons. The business operators of public transportation are not obliged to follow the guidelines; however, it is deemed desirable to use them as a model in conducting the maintenance and improvement of facilities

The target of the guidelines are the passenger facilities which are specified by the New Barrier-Free Act, and this includes railway stations, truck stations or stops, bus terminals, passenger-ship terminals and air terminals. The main user-targets for the maintenance and improvements conducted according to the guidelines are elderly and physically disabled persons and other people whose movements are limited. These guidelines also take in the concept of universal design, so that passenger facilities are easy to use and access is available for all users.

Historical background of universal design

Universal design is a concept which involves designing usable as well as accessible facilities, products and information for people regardless of their various cultures, languages, age, sex, difficulties and abilities. The concept of universal design arose following the congressional passage of the Americans with Disability Act (ADA), which drew much social attention in America at the beginning of the 1980s.

The following seven principles were introduced by Ronald L. Mace in the 1980s.

(1) Fairness
Usable by anybody
(2) Flexibility
Having a wide range of usage
(3) Simplicity
Having a simple structure and being easy to use
(4) Clarity
Necessary information can be understood easily
(5) Safety
Little risk in the event of misuse

(6) Continuity
Usable for a long period of time without feeling mental or physical fatigue
(7) Spatiality
Having enough space for users to feel comfortable

The concepts of "barrier-free" and "universal design" seem to be the same, but their natures are different. The barrier-free concept aims to create situations where physical and psychological obstacles—barriers which cause problems for elderly and physically disabled persons when they participate in social life—are removed according to the related governmental or municipal policies. Thus, it is a form of design which seeks to resolve the problems of users who face difficulties or impossible situations in their daily lives. Universal design is a creative and suggestive form of design. This simulates physical abilities, the psychology and the usage environments of users in advance, regardless of the cultures, languages, ages, sex, difficulties and the abilities of those users. For instance, rigging up a lift to help persons in wheelchairs get onto a bus more easily is an initiative following the barrier-free idea, while lowering the level of the floor of a bus or removing uneven surfaces at bus stops or at terminals so that everyone finds them comfortable to use is following the idea of universal design.

Services and Supports for Persons with Disabilities Act

The Services and Supports for Persons with Disabilities Act was implemented in 2005 in Japan. The purposes of the act are: to contribute to the actualisation of communities and societies where persons and children with disabilities can have independent daily and social lives by receiving necessary benefits and support in accordance with the welfare service for persons with disabilities stipulated by the government; and to realise environments where persons with disabilities can live in peace while having mutual respect for the individuality of others regardless of their difficulties.

Related systems are integrated so that the necessary welfare services are used by persons with difficulties regardless of the types of difficulties they may have, and municipal governments are responsible to offer such services in an integrated manner. The users of the welfare services are to bear some costs (an upper limit is set) of the services provided according to their usage and income, and basically 10 percent of the cost is to be paid by each user. Costs of food and utilities will be also billed to the users of services, and again basically 10 percent of healthcare cost is to be borne by each user.

In addition, in an effort to support persons with difficulties who work at home, special adjustment payments will be given to companies which can pass business orders to people in this category A new job-coaching subsidy system and some job assistance model projects for persons with difficulties—of which the aim is to support them to stay at one job or office for a longer period of time—have been established. With help from these systems in local communities, welfare facilities can supply effective support to persons with difficulties in their working environments. Whether people are eligible to receive the welfare services for persons with disabilities or not is decided by municipal governments which conduct comprehensive reviews in each case.

However, the system of services described above has resulted in imposing a tremendous burden on the lives of persons with disabilities, in spite of the fact that the systems were established under the pretext of supporting their self-reliance. One of the purposes of the system is not so much to protect persons with disabilities as to support their independence. However, the recipients of the welfare service are expected to pay basically 10 percent of the costs of the service. Although the employment of persons with disabilities has been promoted, it is still difficult for them to find jobs. Furthermore, the severer the degree of disability becomes, the more assistance is required. Therefore, it is necessary to introduce policies to reduce the financial burden involved. The problem is that, regardless of any guarantee of income resulting from self-reliance, the expenses involved are expected be paid by the disabled.

ADL and IADL

ADL stands for Activities of Daily Living. This includes basic activities, such as eating, bathing, moving, waking, going to the toilet and other activities necessary for people to conduct their daily lives. ADL was first developed in the rehabilitation sector in order to measure the effectiveness of rehabilitation for patients with functional impairments. Recently, ADL is often applied as a gauge to measure the independence of elderly persons. In this case, the basic life-functions of the elderly are evaluated in terms of three steps, "independence", "part-care" and "full-care", and points are given according to the results. The higher the total score, the more independent a person is judged as being.

IADL stands for Instrumental Activity of Daily Living, and it is based on ADL. IADL measures such things as using the telephone, going out for a long distance, shopping, preparing meals and managing medicine and money. The purpose is to determine whether people have the necessary abilities to have an independent social life. When reviews are to be conducted on the possibility of the elderly or persons with disabilities living at home, evaluations made only according to ADL are considered not sufficiently adequate and IADL becomes an important index for evaluations.

The long-term care insurance system, for example, uses the results of reviews which are conducted based on ADL and IADL, and these results are used as an index to determine the level of care needed by the people involved. The judgements and reviews are made by care managers who are expected to conduct them objectively. In addition, the final decision is made by the Certification Committee for Long-Term Care Insurance organised by municipal governments.

Quality of Life

Quality of Life (QOL) is about people's quality of life from the social point of view, and this is a concept which entails viewing people's lives according to the criteria of how satisfied they are and how happy they are with their lives. This concept was established in 1989 by the World Health Organisation (WHO). It was a proposal regarding the sort of medical care which would place importance on the QOL of cancer patients in all phases of treatment, from the beginning of diagnosis to the terminal phase. Originally, it was used as a gauge in support of the activities of cancer patients. However, QOL has been interpreted over a wider range, and now the concept is also applied to problems relating to, for example, the psychological and social affluence of people, people's mentalities in general, the motivation of elderly people in life, and housing and other environmental issues.

In the field of medical rehabilitation and welfare, the goals had only been the movement-recovery of patients and the supply of care toward activities related to patients' daily affairs (ADL) until the 1960s. However, in the 1970s, campaigns for Independent Living (IL) were carried out in America. In these campaigns, various people including persons with disabilities stressed the idea that if persons with disabilities could receive care from other people, then they could participate in society and could have some form of self-actualisation even if they were unable to take care of their own daily affairs by themselves. This movement had a great impact on the Rehabilitation Act implemented in 1978 in America, and this produced a turning point in the aims of rehabilitation. Rehabilitation changed from the improvement of ADL to the improvement of QOL. The idea of placing an emphasis on QOL has spread to the world of nursing care, and the mainstream of the concept is now improving QOL rather than ADL.

Rehabilitation

Rehabilitation is a comprehensive approach which involves returning persons with disabilities to more active daily life by helping them recover their abilities as much as possible and by improving their independence. Terms such as "returning to society" or "treatment and education" were used in place of today's term "rehabilitation" originally in Japan, but the term "rehabilitation" has become more common since around 1960. The term "rehabilitation" is still understood as physical training for functional restoration in Japan, though the original meaning of the term is the recovery of the rights of persons with disabilities so that they might live in a better manner. Training for functional restoration is nothing more than an avenue to achieve the greater purposes of rehabilitation [(8)].

WHO defined the term rehabilitation in 1981, and it stated that the term includes every kind of treatment and procedure of which the aims are to reduce the negative impact from situations which could cause learning impairments or social disadvantages to persons with disabilities, and also to help persons with disabilities to achieve social integration. According to Satoshi Ueda [(9)], the goal of rehabilitation is not only to help the persons with disabilities adjust to their environment through training, but also to make their social integration easier by intervening directly in their environments and the entire society. In addition, persons with disabilities, their families and their communities must get involved with plans and operations related to various kinds of services connected to rehabilitation.

Moreover, the preservation of health, treatment and other aspects related to medical care, which are conducted by specialists in medicine as well as specialists in related fields, are called medical rehabilitation, and this sometimes includes areas which provide treatments and training such as physiotherapy, speech therapy, the making of artificial limbs, occupational therapy and psychological training.

The list below details the types of specialists and their jobs.

(1) Physical Therapist (PT)
Physiotherapists support the independent daily life of their patients by providing exercise therapy and physical therapy, of which the aims are the recovery and the maintenance of basic motion abilities of patients, as well as the prevention of the deterioration of disabilities.

(2) Occupational Therapist (OT)
Occupational therapists support and promote their patients by training them in crafts, such as carpentry. Their aim is to help their patients return to their social lives.

(3) Speech Therapist (ST)
Speech therapists provide training and support to persons with speech and hearing impairments, and the aim is to promote their functional recovery and improvement.

(4) Clinical Psychologist (CP)
Clinical psychologists help patients who suffer from mental illness, psychosomatic illness or psychological problems, and the aim is to try to maintain the mental health of the patients by using clinical psychology.

(5) Medical Social Worker (MSW)
Medical social workers are social workers in the field of healthcare, mainly in

hospitals. They provide help in solving and coordinating psychological and social problems suffered by patients and their families from the standpoint of social welfare.

Psychological impact on persons with congenital disabilities

The impact of impairments on the life and behaviour of persons with disabilities differs depending on the types, degrees and timing of the occurrence of the impairments. When persons with congenital disabilities are still infants, the biggest problem is usually found in the minds of their parents. The parents usually suffer from shock, a guilty conscience or anger. When persons with congenital disabilities reach childhood, a sense of "self" develops and they become conscious of their own disabilities. Then, when the centre of their life shifts from home to school, they take the differences between themselves and other people very hard and suffer from shock. During puberty, they try to establish distinctiveness, and they have difficulties in accepting their disabilities while the limits on their daily activities increase their stress-levels.

In adulthood, persons with congenital disabilities try to be independent through social activities. They experience incomprehension and prejudice in society, and they feel conflicted in themselves. However, in their 40s, they begin reconciling themselves to their disabilities and start to reach a psychological settlement. Entering old age, they become more conscious of a lack of physical strength than of any psychological problems relating to their disabilities. They become favourable towards communicating with other persons with similar disabilities and they have a strong tendency to depend on their peer group.

Acceptance of disability by people who acquired disabilities and developmental disabilities

Ueda (10) describes the acceptance of disability as neither a matter of giving up nor of being defiant of one's condition, but rather a matter of persons attaining a transformation of values about their disabilities. This involves them overcoming any feeling of shame or inferiority about their condition and changing their attitude in life to a more positive state by realising and understanding that having a disability does not lower their value as a human being. In addition, Ueda describes the acceptance of disability by people through a merging of the traditional theory of transformation of value with stage theory. The details are explained below.

(1) Shock period

This is a psychological state just after the development of disability (the onset of disease, or an injury). Patients at this stage feel psychologically peaceful and they are in the state of apathia and accidia, though the disability may be physically painful. They feel little anxiety and have a similar level of purpose and desire in their daily lives to the prior condition when they were healthy. They can communicate with other people in a manner similar to before the development of disability.

(2) Denial period

As a disabled person's physical condition settles, their biologically protective reaction disappears. They begin to realise that their disability will not be healed easily. Denial of disease or disability begins as a psychologically protective reaction. Denial is divided into two types; conscious denial and subconscious denial. Though subconscious denial is not easily observed, it has a bigger impact on the behaviour of patients than conscious denial.

(3) Period of confusion

This is a period which develops in consequence of patients having no effective ways to deny their sense of reality and the impossibility of their disability being healed completely. Patients blame everybody for everything, and take out their anger on everybody. When the emotion appears in an introverted or self-accusatory way, they begin to blame themselves. They blame themselves for everything, and they become depressive.

(4) Period of effort towards a resolution

This is a period of making positive and constructive efforts. Patients begin to realise that their outward attacks are not a way to solve their issues. They understand they must realise that their disability is their own responsibility and that they have to make efforts by themselves without relying on other people. As a precondition to their efforts, it is essential for them to have a bright and realistic vision, and foresight such as seeking the improvement of motion-ability in daily life, a chance of returning to work, or a reduction of social disadvantages.

(5) Period of acceptance

The transformation of value is completed gradually, and patients begin to take actions towards finding new roles or jobs in their society, and they start once more to find their life worth living.

CHAPTER TWO

Physical and mental features of persons with a limb/trunk dysfunction

Definition of the term "disability" in Japan

It is commonly misunderstood that "disability" means a condition where the state of an illness or/and injury is fixed and will not heal. However, the term "disability" actually refers to the difficulties, inconvenience and disadvantages which persons face as a result of an illness or/and injury. The term "person with a disability" is defined in Article 2 of the Japanese Basic Act for Disabled Persons as a person who receives a considerable degree of limitation in his/her daily life or his/her social life for a long period of time. By contrast, Article 4 of the Physically Disabled Persons Welfare Act defines a person with a physical disability as a person who is over 18 years old with physical disabilities and with a physical-disability certificate issued by his/her prefectural governor.

The International Classification of Impairments, Disabilities and Handicaps (ICIDH) proposed by WHO in 1980 defines physical disability in a three-level structure, consisting of: Impairment, Disabilities and Handicaps [(11)]. The first level is impairment. Impairment is understood to be the primary level of disabilities which have directly developed from disorders, and thus this classification is based on the biological level. This includes paralysis, anchylosis, decline in physical strength and alogia. The second level is disability. Disability is understood to be the deterioration of ability which is caused by impairment, and thus this classification is based on individuals. The third level is handicap. Handicap is understood as the condition whereby people receive social disadvantages as a result of disability, and this classification is thus based on humanity. People's disabilities may become more severe depending on their social circumstances.

Akira Shimazu [12] illustrates how to provide rehabilitation to persons with disabilities, as is described below. Patients at the first level are subject to treatments which, for example, promote functional recovery and prevention and the medical treatment of secondary complications. Patients at the second level are subject to compensatory treatments, which target their functional disabilities. These compensatory treatments include training to improve the strength of the healthy parts of a limb, the supply of artificial limbs, necessary orthosis self-help tools and wheelchairs, education in new methods to improve patients' activities in daily living and skill training to improve their social lives and relationships with other people. Patients at the third level are subject to improvements to their environment. These improvements can be made by remodelling the patients' house, teaching how they can assist themselves in order to be independent, promoting a return to work, introducing hobbies and sports to improve the quality of their lives and providing them with indemnity, such as educational or welfare-income indemnity.

Furthermore, a psychological approach is important to deal with the sense of psychological deprivation among patients. It is important for us to realise that a society becomes normal when persons with disabilities share their lives with other people rather than being eliminated from that society. This is the valued concept called the normalisation of society, which was described earlier.

Physical features of persons with limb/trunk dysfunctions

People who have permanent motor-functional disabilities in all or part of their four limbs (upper limbs and lower limbs) and body trunk are called persons with a limb/trunk dysfunction. Motor-functional disability refers to the condition of a person whereby the functions of parts of his/her body deteriorate or stop and, as a result, he/she has difficulties or disabilities with his/her movements. The parts of the body which relate to movement include the nerve, the muscular, the osseous and arthrosis systems. The function of the nerve system is to receive and convey information from and to every part of body. The human body moves when the muscular, osseous and arthrosis systems operate according to information received from the nerve system. The condition whereby the nerve or/and muscle system stop functioning is called "paralysis." Disabilities with motor function are thus called "motor paralysis", and disabilities with senses of touch, temperature, pain and other sensory functions are called "sensory paralysis." According to the classification of regions of the body, paralysis in both upper and lower limbs is called "quadriplegia", paralysis in either of the upper limbs or lower limbs is called "paraplegia" and paralysis on either the right or left side of the body is called "hemiplegia."

There are seven features of disabilities with motor functions, and the conditions can be described as follows:

(1) Abnormality of muscle tension
As a result of abnormality in the peripheral nerve and the central nerve, muscle tension becomes either deteriorated or accelerated. As muscle tension relaxes, the muscle thins down and becomes weak, and sometimes the scope of joint movements becomes abnormally large. When muscle tension accelerates, the arms and legs become too stiff to make movements. When pressure is added to the arms of a person for example, he/she has difficulty at first in moving his/her arms up to a certain angle, and then he/she suddenly loses strength and resistance, or a certain level of resistance in the arms continues.

(2) Weakening of muscle
When abnormality in muscle tension or the condition of not using a muscle continues, the muscle becomes thin and weak.

(3) Reduction in the scope of joint movements and deterioration of movement
When the anatomy, muscle and ligaments around arthrosis become abnormal, the scope of joint-movements becomes small. Abnormalities in the bones which make up joints or of the cartilage in joints also cause a reduction of the scope of movement, as well as deterioration in the smoothness of the movement of joints. Sometimes, the joints become stiff at certain angles and this prevents joints from moving properly or at all.

(4) Loss of the whole or part of leg(s) and/or arm(s)
There are two types to this condition. One is the case where a part or all of the four limbs is/are missing due to congenital causes. The other type is the case where a part or all of the four limbs is/are missing because of being cut or amputated due to accident or illness.

(5) Body movements against the patient's will
Disabilities and damage in the development of the nerve system of the brain can cause motor activities against a patient's will.

(6) Patients are incapable of making coordinated motor activities or fine motor activities
When patients try to move, their arms and/or legs may only swing or dangle. Patients find they have difficulty in making fine or skilled motor actions, such as using chopsticks and writing, with their arms and/or legs.

(7) Changes in the shape of the whole or of parts of the body
Changes in the body can be caused by muscle shrinkage, bone malformation and the stiffening and hardening of joints.

Psychological features of persons with limb/trunk dysfunctions

Patients who have acquired or developed disabilities with their motor functions sometimes find it difficult to accept their disability. There are some cases of patients being too obsessed with their own recovery, so that their social rehabilitation is very difficult. Sometimes, the families of patients react in the same way. However, the acceptance of disability can become easier for patients if they succeed in finding what they can do, making images of a new life adjusted to their disabilities and finding purpose in their lives through new roles and activities among their family and society.

Spinal-cord injury

The bundle of sensory nerves and motor nerves in the spinal column is called the spinal cord. The spinal nerves emerge from the spinal column, and they branch out symmetrically and extend to the peripheral parts of the body. As

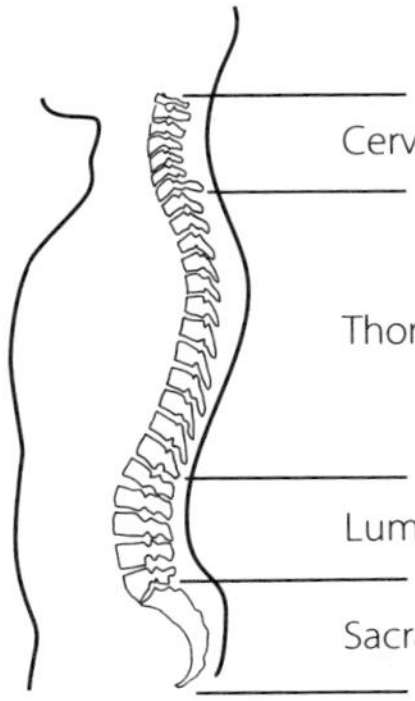

Figure 2-1 Spinal cord nerve segments

shown in Figure 2-1, the names of the spinal nerves are abbreviated as: C for the cervical nerves, T for the thoracic nerves, L for the lumbar nerves and S for the sacral nerves. Spinal nerves comprise of 8 pairs of the cervical nerves, 12 pairs of the thoracic nerves, 5 pairs of the lumbar nerves, 5 pairs of the sacral nerves, as well as the coccygeal nerve.

Spinal-cord injury is often caused by strong external pressure on the spine, and impairments of motor functions, sensory functions and autonomic nerves are often caused by damage to the spinal cord, which makes up the central nervous system together with the brain. The spinal nerves are responsible for a wide range of motor and sensory functions in the human body; thus impairments of the spinal cord tend to be very severe. Cervical-cord injury is the most serious injury of the spinal cord, and its damage causes quadriplegia and respiratory impairment in many cases.

Damage to the area lower than the mid-chest area of the spine will not damage upper limbs but results in paraplegia. Spinal cord injury impairs motor function and, simultaneously, it triggers functional impairments of the senses, the body-trunk, autonomic nerves and excretory organs. The symptoms of these impairments are as described below.

(1) Impaired motor function

The symptoms of impairment of motor function include paralysis and spas-

Table 2-1 Spinal-cord injury levels and ADL (activities of daily living)

Injury levels	Motor functions	Mobility and movement
C1-C3	• Limited head and neck movement. • Impairment of respiratory system. • Complete paralysis of arms, legs and body.	• Electric wheelchair which is remodelled for special needs.
C4	• Able to breathe without assistance. • Able to lift scapula.	• Specially remodelled electric wheelchair.
C5	• Able to move parts of shoulders, elbows and forearms.	• A manual wheelchair with hand-rims on flat surfaces.
C6	• Incomplete but good shoulder movement. • Full elbow flexion but no extension. • Able to do a few push-ups . • Able to flex the wrist, though it is weak.	• A manual and practical wheelchair adapted with hand-rims on flat surfaces. • Able to transfer from bed to wheelchair and from wheelchair to bed with or without assistance. • Remodelled car. Assistance may be required.
C7	• Complete movement of hand up to the wrist. • Full wrist extension and flexion. • Able to do push-ups.	• Standard manual wheelchair. • Able to transfer from bed to wheelchair and from wheelchair to bed independently. • Able to transfer from toilet bowl to wheelchair. • Remodelled car.
C8-T1	• Full use of upper limbs.	• May use a manual wheelchair on uneven surfaces. • Remodelled car. Able to load wheelchair into car without assistance.
T2-T6	• Semi-body-trunk balance.	–
T7-L2	• Nearly complete body-trunk balance. • Able to elevate the pelvic girdle.	• Able to walk with orthosis (though poorly in practice) and crutches.
L3-L4	• Full body-trunk balance. • Able to move parts of lower limbs.	• Able to walk with ankle-foot orthosis.
L5-S3	• Incomplete movement of ankle joint.	• Able to walk independently.

Source: Kan Nomura: Textbook for 2nd/3rd Welfare Living Environment Coordinator, published by Tokyo Chamber of Commerce and Industry, 2001.

ticity. If the paralysed leg of a person with spasticity is touched or moved, for example, his/her leg may become spastic without any relation to his/her will. The symptoms become worse in a cold environment.

(2) Impaired sensory function

A person with impairments to their sensory functions has complete numbness to pain, heat, coldness and other senses relating to temperature, with the parts of the body that are paralysed. This condition means delays in detecting injuries or burns. In addition, persons with impairment to their sensory functions tend to suffer from decubitus. Decubitus is caused by bad blood circulation in regions of the body, and occurs because these regions are being put under pressure for a long period of time. As a result, the skin and flesh of the parts concerned become necrotic. If the condition gets worse, the area will be infected and sometimes this even results in death. In order to prevent decubitus, the position of a patient's body should be changed frequently.

(3) Impaired trunk function

A person with impairment in trunk function will have paralysis of the abdominal muscles, muscles of the back and other muscles, and thus it is very difficult to maintain a normal sitting posture. In particular, if the injury is at a higher level than C6, the person with impairment in trunk function will fall over unless his/her body is fixed to, for example, a chair.

(4) Impaired autonomic nerves

A person with impairment to the autonomic nerves has difficulty in controlling body temperature because he/she does not perspire. Moreover, placing the body in a sitting position may cause anaemia. When urine is collected in his/her bladder or he/she wants to empty his/her bowels, blood pressure rises suddenly and this can cause headaches, diaphoresis and the worsening of spasticity. Deterioration of these symptoms may cause a brain haemorrhage and create a critical condition for the person.

(5) Impaired excretory function

Paralysis of the muscles used for egestion prevents a patient from regular egestion. Thus, laxatives and suppositories are used according to scheduled dates in order to empty the patient's bowels, and a tube is inserted into the bladder to discharge urine.

Progressive disorders

Common progressive disorders involving motor-functional impairment are as described below.

(1) Progressive muscular dystrophy
Malnutrition of muscular tissue weakens muscles and stops the whole body or upper and lower limbs from moving. It is a genetic disorder, and there is no effective treatment except prevention.

(2) Duchenne muscular dystrophy
Most progressive muscular dystrophy belongs to this type, and the frequency of onset is very high. If it develops in early childhood, the patient will stop walking at around 10 years old, and he/she will be able to move about only in a sitting position or by using a wheelchair. Around age 15, the patient will have difficulty maintaining a sitting position, and a manual or electric wheelchair will be necessary for transferral. The muscles of the upper limbs also weaken; however, strength will be kept up to a certain level. The patient will survive until around 30 years of age today due to improvements of treatment with artificial respirators. Duchenne muscular dystrophy is a sex-linked recessive genetic disorder, and thus it develops only among males.

(3) Limb girdle muscular dystrophy
The muscles of the upper and lower limbs weaken and become thinner. Both males and females can be affected with this, and in many cases it develops between the ages of 10 and 20, though some develop it as late as between the ages of 30 to 50. Muscle tissue weakens from the pelvicrural to the shoulder girdle muscles, or in the opposite direction, and amyotrophy develops. Muscle sudohypertrophy also develops sometimes. It develops slowly, and patients will have difficulty in walking within several years. Though the degree of seriousness or the speed of development varies greatly depending on the person, generally the patients will have a high degree of functional impairment when they become middle aged or older.

(4) Facioscapulohumeral muscular dystrophy

This is an autosomal dominant inheritance disorder. It can develop sometime between childhood and adulthood. The disorder develops first with amyotrophy and muscle loss in the face, shoulder girdle and brachial region, and then it develops in the pelvicrural muscle. The muscles of the pelvic girdle also weaken in later years. It develops very slowly, and so patients can live for a long time.

(5) Spinocerebellar degeneration

Ataxy is a diagnostic of this disorder, and patients have difficulty with little movements. The nerve cells between the cerebellum and the brain stem, as well as the spinal cord, are destroyed or disappear. Patients' bodies move sideways when they walk. The disease develops chronically, and when it becomes severe the patients must be confined to bed. The disorder was designated "a project for research into treatment for a specific disease" by the Health, Labour and Welfare Ministry in 1976.

(6) Amyotrophic lateral sclerosis

Symptoms of weakening or the thinning down of muscles develop in some parts of body, depending on the type of the disorder. Types of the disorder are categorised into: a type with weakness or stiffness of hands or arms; a type with drop foot; and a type with difficulties of vocalization or speech, or the swallowing of food and drink. It develops in people between 40 and 50 in many cases, and it develops more among men than women. It develops slowly, and almost all of the motor functions, except the muscles for eye movement or some others, will be destroyed. Many of the people affected need to have an artificial respirator because the respiratory muscles are paralysed. Half of the patients die 3 to 5 years after the outset of the disorder, and effective treatments have not been established yet.

Cerebral paralysis

Brain damage received in the period between conception and 4 weeks after birth can cause palsy, paralysis or other nerve impairments. The diagnostic of this disorder is a condition in which the patients have abnormalities with their muscles, and this prevents the patients from moving the muscles in their four limbs and body trunk.

The common types of this disorder are as described below.

(1) Muscle-spasticity-type cerebral palsy
Anchylosis of the four limbs is one of the diagnostics for this. The pattern is categorised in hemiplegia, paraplegia, quadriplegia and diplegia, depending on the region with impairment. As muscle tension accelerates, the arms and legs become too stiff to move according to the will of the patient. Once body posture becomes off-balance, it is hard to regain the desired position. The degree of impairment can be categorised from slight to severe.
(2) Athetoid-type cerebral palsy
Patients have damage to the basal ganglia, and its diagnostic is involuntary movement. The body moves involuntarily. Muscle tension becomes both extremely high and low. Unexpected and unregulated movements are seen, such as sudden movement of the head or stiffness in the arms. Speech impairment is also found in many cases, and the impairment can be categorised from slight to severe. The number of patients with athetoid-type cerebral palsy accounts for a quarter of all patients with cerebral paralysis
(3) Mixed-type cerebral paralysis
This is a type in which muscle-spasticity-type cerebral palsy and athetoid-type cerebral palsy are mixed. Patients suffering this type have diagnostics of anchylosis and unexpected movements, and they suffer from quadriplegia in many cases. The athetoid-type used to be common among this type, but the muscle-spasticity type has become more common in recent years due to advances in medical technology.
(4) Ataxic-type cerebral paralysis
The cerebellum or the pathway of the cerebellum of patients is damaged, and the weakening of muscle tissue is one of the diagnostics for this. The muscle tension of the patients is low, and thus the patients often have difficulty with picking up objects, and their movements are often unstable.

Amputees and artificial parts of the body

Amputees are people with disabilities of motor function because they have lost, or had amputated, part or all of their four limbs, as a result of injury, disorder or congenital malformation. There are many cases of upper-limb amputation. In a number of cases, patients with certain illnesses have upper limbs amputated because of bad blood circulation at the distal portion of extremities. After the amputation of legs or arms, the patients may feel as if the leg or arm still exists. This is called a "phantom limb." Patients who have a phantom limb may move it intentionally. However, patients who cannot move the phantom limb sometimes feel strong pain in the region which does not actually exist, and this is known as "phantom limb pain."

Figure 2-2 shows some replicated parts of the body. This replication helps patients to restore parts of body lost due to congenital causes, accidents or sickness, without having an operation. This is different from using artificial hands or legs, which focus on functionality. For example, the colour of the skin and the wrinkles of the users are replicated precisely, and the users can take a bath or go to a pool, for example, while wearing the replica or attaching it to their body. The material is made of durable silicon, which is especially developed for human use, and special make-up techniques are also applied to create the replicas. The user feels no pain from wearing or attaching it to their body, and repairs are also possible.

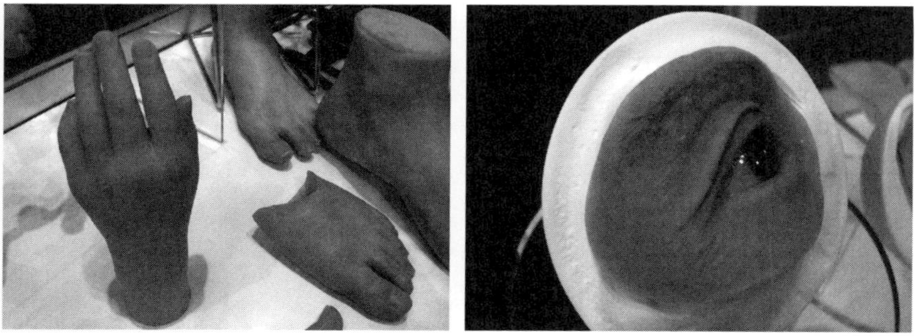

Figure 2-2 Order-made artificial parts of a body (Kawamura Gishi)

Disuse syndrome

One part of disuse syndrome is decubitus, which was described earlier. This syndrome is related to a circulatory deficit of the skin. Decubitus is very often found among patients who use wheelchairs or who are confined to bed. The common causes of decubitus are: pressure (a defect in the body pressure balancing function, caused by a circulatory deficit); moisture (diaphoresis, incontinence); rubbing; and internal causes (malnutrition, old age, swelling, sensory impairment, depression of body-motion ability). Some of the factors coexist and cause decubitus to develop. The warm, humid climate in Japan is closely related to the country's high incidence of decubitus.

Persons with spinal-cord injuries can feel no pain in the region of paralysis. Moreover, when they are in wheelchairs, decubitus may develop because they cannot feel the pressure on the skin around the buttocks. Therefore, it is neces-

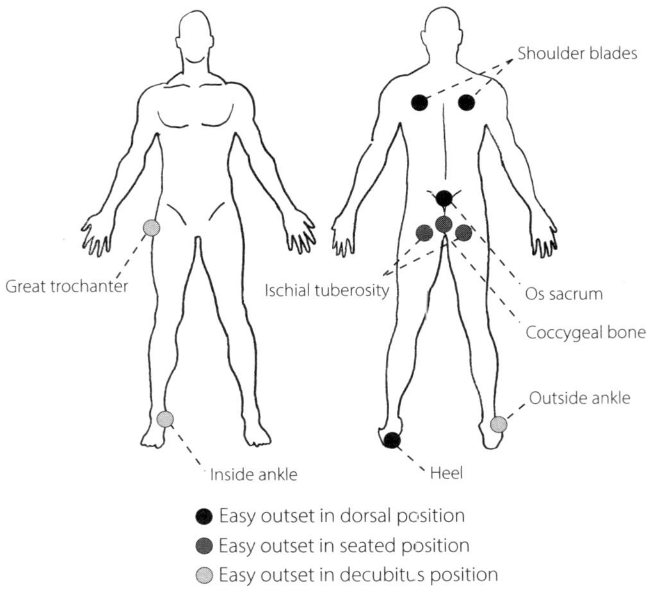

Figure 2-3 The regions where decubitus appears most commonly in the human body

sary to lift the buttocks or to move the body sideways at least once every hour, in order to help the flow of blood to the skin around the buttocks. The regions where decubitus appears most commonly among persons with damage to the spinal cord are shown in Figure 2-3. These areas tend to have high pressure on the area between the bone and skin.

Nourishment, joint deformity and depression of motor ability can also bring on the condition of decubitus. Thus, the treatment of decubitus requires various approaches. The use of auxiliary tools can also help prevent decubitus from appearing.

CHAPTER THREE

Wheelchairs and barrier-free sidewalks

The history of the development of wheelchairs

The history of wheelchairs goes back a long way. Shown in Figure 3-1 is a picture of a bed with wheels which was used in Greece around 530 B.C., while shown in Figure 3-2 is a picture of a wheelchair with spokes which was drawn around 525 A.D. in China. The wheelchair depicted in Figure 3-2 is allegedly the

Figure 3-1 A picture of a bed with wheels painted on a vase
Source: http://www.wheelchairnet.org/wcn_wcu/slidelectures/sawatzky/wc_history.html

Figure 3-2 A picture of a wheelchair drawn on a slate
Source: http://www.wheelchairnet.org/wcn_wcu/slidelectures/sawatzky/wc_history.html

oldest recorded image of a wheelchair.

It is said that a chair with wheels (see Figure 3-3) was first built when King Philip II of Spain became ill in Europe in 1595. The first self-propelled wheelchair was designed by Stefan Farfler, a clock-smith, in Germany around 1655 (see Figure 3-4). In the 18th century, components of wheelchairs were developed. Drive wheels were replaced with caster wheels and a removable foot plate was often equipped to a wheelchair. In addition, some wheelchairs were decorated with accessories to resemble furniture.

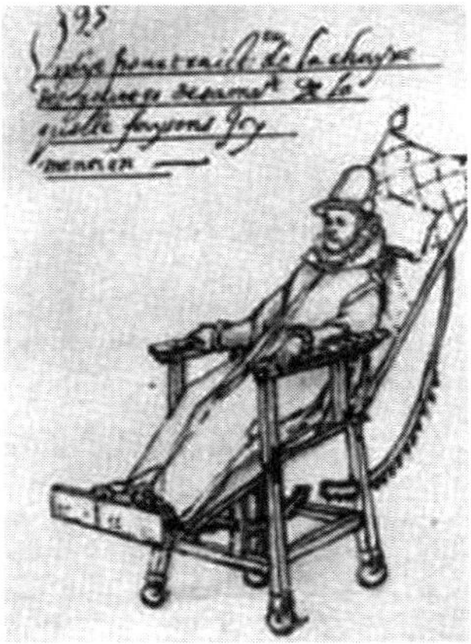

Figure 3-3 The wheelchair used by Philip II
Source: http://www.wheelchairnet.org/wcn_wcu/slidelectures/sawatzky/wc_history.html

Figure 3-4 A wheelchair which was designed to be propelled by hand
Source: http://www.wheelchairnet.org/wcn_wcu/slidelectures/sawatzky/wc_history.html

In the 19th century, chain-driven wheelchairs were designed. Light wheelchairs were produced at a cheaper cost, and it became more common for the occupants to control the wheelchair by themselves. In 1914, the first ever folding wheelchair was built experimentally, and a powered wheelchair with a full weight of 135 kg was produced in 1915. Folding wheelchairs were commercialised in 1933. Against such a background, wheelchairs were developed, and then they were mass-produced and widely used by the wounded veterans of World War II.

Around 1915, wheelchairs were imported from America and England to Japan. However, these wheelchairs did not fit with the configuration, style and size of Japanese houses. In addition, the imported wheelchairs were generally too big for Japanese people. Consequently, the production of wheelchairs began in Japan. The first wheelchair made in Japan was called a "Kairin-jidousha", which can be translated as "revolving auto-wheels". It is said that this first wheelchair was produced by a company which made rickshaws.

One of the oldest wheelchairs of which detailed records exist in Japan is a "wicker-work wheelchair" which was produced by Tojiro Kitajima in around 1920. He was the founder of Kitajima Shokai which is now K&AI Corporation. After the war, Kitajima replaced the wicker part of the chair with cloth and he developed a three-wheeler wheelchair, which had two front wheels and a rear wheel which was driven by a chain. This wheelchair was called an "autokinetic car." The autokinetic cars were mostly used in the Hakone Clinic for Disabled

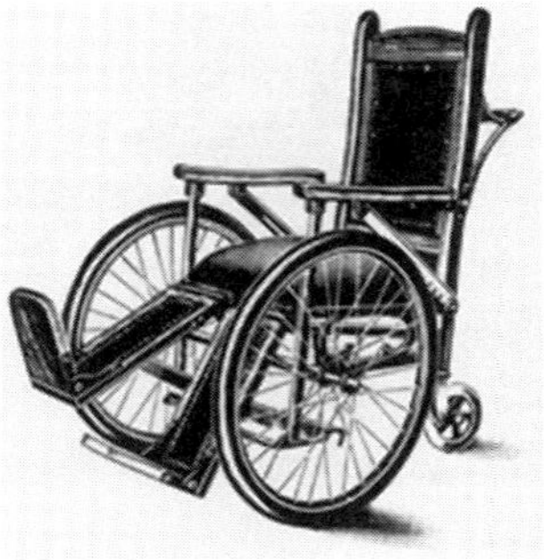

Figure 3-5 The Hakone-style wheelchair
Source: K&AI Corporation

Veterans in Japan, which was the predecessor of the National Hospital Organization Hakone National Hospital. Because of this, the wheelchair has also been known as a "Hakone-style wheelchair" (see Figure 3-5).

In November, 1964, the Paralympics was held in Tokyo, Japan, and this fostered a greater momentum in the production of wheelchairs. Study of wheelchairs started in full swing in medical and engineering circles. The outcomes of this study were put together as "a projection of wheelchairs for spinal-cord-injury victims." In 1971, Japanese Industrial Standards (JIS) for manual wheelchairs were established, based on the projection (see section "Electric wheelchairs and Japanese Industrial Standards" for more details). Since then, improvements have been made to materials and the modularisation of wheelchairs, and adjustable wheelchairs have also been developed.

The first electric wheelchair in Japan was developed in 1968 by the Yaesu Rehabili Company, which was the biggest rehabilitation-device maker between 1960 and 1970 in Japan. Thus, the history of the electric wheelchair is very short in Japan. Sophisticated electric wheelchairs have been imported from overseas to Japan, but the cost and other factors, which are described below, have prevented electric wheelchairs from being used widely in the country. Takashi Masuzawa (13) has pointed out that some factors are related to the exclusive characteristics of the Japanese market for welfare equipment. However, the Japanese culture and lifestyle have had a bigger negative impact on the spread of imported wheelchairs in Japan. Masuzawa mentions that the Japanese climate is hot and humid, and therefore Japanese houses are built with high-floors. Furthermore, Japan has a rainy season. Thus, Japanese people take off their shoes when they enter houses and some buildings, in order to prevent dirt from being brought inside. This has created a clear line between the inside and outside in the minds of Japanese people. Traditionally, Japanese houses had no corridor, and the rooms or living spaces were divided by such as paper and folding screens. Since westernisation from the Meiji period (1868-1912) onwards, many western-style houses have been built. In such houses, rooms are divided by walls and people enter or leave the rooms through doors. However, houses are still built based on the units called "shaku" (1 shaku = about 30.3 cm) and "ken" (1 ken = about 1.82 m) in Japan. Consequently, the general structure of Japanese houses is still very different from America, Europe and other countries.

Under such circumstances, the use of imported wheelchairs is difficult for Japanese people.

Design and product configuration

The design concept of manual wheelchairs is very different between Japanese and Western products, and their product configuration also differs a great deal. Generally, Japanese wheelchairs are built by means of weld assembly. As a result, once the wheelchair is built, repairs and adjustments of, for example, the width or depth of the seat or the height of the armrests are difficult to manage. In contrast, module-model wheelchairs are common among western wheelchairs. The body frame, wheels, seats and other components are standardised in accordance with a common design concept. Therefore, the recombination of components of a wheelchair can be done freely, and accordingly, the variation of sizes, colours and models is abundant. Even after the completion of assembly, adjustments or changes of components are possible for each wheelchair in accordance with each user's body size and level of disability, and also with wear and failure of components and other factors. However, the familiarisation of module-model wheelchairs in Japan demands the training of specialists and engineers who have both specialized knowledge and skills as required, according to a study of the welfare systems in foreign countries conducted by JETRO (Japan External Trade Organization) [14].

Regarding electric wheelchairs, Japanese electric wheelchairs (with a joystick steering control) afford no structural adjustment, and the range of program-settings of steering and speed control is limited. In contrast, western electric wheelchairs afford detailed configurations of programs, and the configurations and adjustments are flexible in accordance with the level of disability, body strength and other needs of the users.

The body size and gender of users are key points in designing a manoeuvring device for wheelchairs. For example, the position and size of the manoeuvring device are generally determined based on a mean value of people's size. However, if the decision is made based only on the mean value, some people, such as people with longer or shorter arms than normal, usually find it difficult to operate their wheelchairs comfortably. The design of seats for wheelchairs is another example. The seats of wheelchairs are designed to hold and keep the body trunk of the occupant in an upright position. Thus, if the body of the occupant moves every time that he/she controls the wheelchair, then the design specifically made to keep the body trunk in its upright position loses its positive effect.

In addition, women are generally shorter than men, but wheelchairs are usually designed based on a mean value which is calculated using the total number of men and women. When sitting on a wheelchair with a high-back-seat, the body trunk of the occupant is likely to drop forward. Moreover, elderly female people in Japan are often very small and they find the depth of seat is too deep for them. In such a case, the occupant is highly likely to suffer from thoracic kyphosis. In addition, according to T. Hettinger [15], muscle power reaches 100 % at the ages of 25-35 and then power declines to only 75-85 % at ages of 50-60, regardless of gender.

Models and configurations of wheelchairs

A legal act for "the research development and the promotion of improvements for welfare equipment" was implemented in 1993 in Japan. It states that the

Table 3-1 Classification of wheelchairs

Manual wheelchair	Basic model	Reclining-model
		Manual lift-model
	Front drive-wheel model	Reclining-model
	One-hand-drive model	Reclining model
	Manual chain-drive model	Reclining model
		Lever-drive model
	Hand-push model	Reclining model
Electric wheelchair		Regular model (4.5) km/h
		Regular model (6.0) km/h
		Reclining-model
		Electric reclining-model
		Electric lift-model

purpose of a wheelchair is to compensate the partial loss of the body or functional impairments of persons with disabilities. Wheelchairs are categorised as adaptive equipment, which is a general title for equipment that is used to ease the daily and occupational lives of persons with disabilities. Wheelchairs are then divided into manual and electric wheelchairs. A variety of wheelchairs are thus being produced today in Japan. According to the standards supplied by the Physically Disabled Persons Welfare Act, wheelchairs are classified as shown in Table 3-1.

The type of a wheelchair is determined by the user according to the level of his/her disability or disabilities. Furthermore, wheelchairs are classified as ready-made models, order-made models and module-models. Figure 3-6 shows the names for parts of a wheelchair.

(1) Ready-made models

Since mass-production of wheelchairs began, steel-made wheelchairs have been sold at low prices even in hardware stores and on the Internet, and

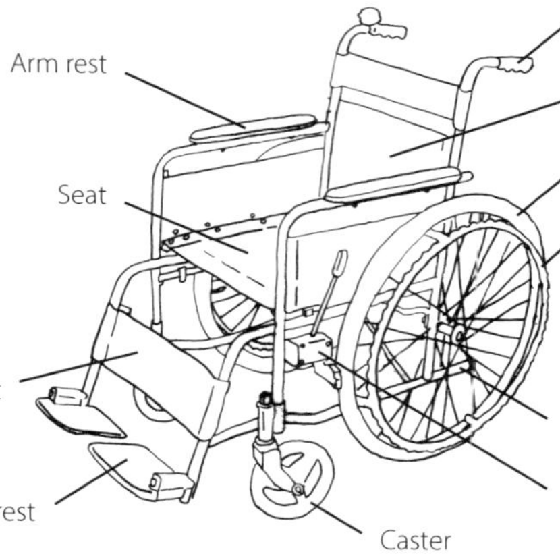

Figure 3-6 Names for parts of a wheelchair

recently it has been possible to buy even aluminium-made wheelchairs at relatively low prices. Steel-made wheelchairs are high in strength, but they are heavier, as some steel-made wheelchairs weigh over 20 kg, while the net weight of an aluminium-made wheelchair is often less than 10 kg. This weight makes it difficult for the disabled to transfer a steel-made wheelchair into a car. In addition, choosing the right wheelchair is difficult, especially for persons with severe physical impairments and whose body trunks are deformed. In the case of ready-made model wheelchairs, the sizes are pre-determined and adjustability is very limited.

(2) Order-made models

Order-made model wheelchairs are produced in accordance with, for example, the size of the body, the range of motion of the four limbs and the intended purpose of the occupant. Though the price tends to be high and a long period of time is required before the finished wheelchair is delivered, the resulting chair fits the user's body very well. However, there are limits to the adjusting mechanism, and thus, if the body structure or symptoms of the users are expected to change over time, a module-model wheelchair is usually more suitable. It is wise to receive advice from physical therapists and occupational therapists over this.

(3) Module-models (semi order-made models)

A module-model wheelchair is assembled with components which are selected by the user based on the user's body size and wishes. Almost all of the components are assembled with bolts, and it is easier to make adjustments if, or when, the body size of the user changes, or if some medical condition causes, for example, a stiffening or deformation of the user's skeleton. However, this model is lower in strength compared to the models without adjustment mechanisms.

Hand rims and wheels

Hand rims are wheels which are attached to the outside of the drive wheels on a wheelchair and they are slightly smaller than the drive wheels. When the hand rims are rolled forward, the wheelchair moves forward, and when they are rolled backward, the wheelchair moves backward. When one hand rim is rolled, the wheelchair turns around in a direction opposite from the direction in which the hand rim was rolled.

The shape and cross-sectional surface of a hand rim varies depending on the maker, and the standard hand rim is a circular tube. The materials also vary, and they include aluminium, titanium and plastics. Hand rims are often very thin, and people who have a disability with their hands find it difficult to grip them. According to Bengt Engström (16), when hand rims are too thin, the user is likely to grip the rim together with the tire (the drive wheel), and this will increase the risk of his/her hand or fingers getting injured. Thus, a hand rim with a minimum of 2.5 cm in diameter is ideal for a better grip. In addition, the surface of the hand rim should not be too slippery, but a hand rim with a no-skid sole may cause burns to the hands of the user when they stop the wheelchair suddenly. Anti-slip gloves can increase the friction between hand rim and hand, and it can also protect the user's hands. Rolling a wheelchair can damage the user's hands and fingers, especially the thumbs. Abrasion, sprains of joints and blisters on the palms are very common among wheelchair users.

The drive wheels are placed at both sides of the frame of a wheelchair. The wheels used for basic wheelchairs are generally similar to the wheels of bicycles. The size of a wheel for a basic, manual wheelchair is about 22-25-inches in diameter. The wheelchair moves forwards or backwards when the occupant grips and rolls the hand rims. The wheelchair can make a right turn, a left turn and half-turns by rolling one wheel. Some wheelchairs are designed and built for a caregiver to control them. Such wheelchairs are built with no hand rims, and the drive wheels are smaller than the wheels of manual wheelchairs. The size is then around 12-16-inches in diameter. There are two types of tires. One is a pneumatic tire, and the other is called a solid tire, which is made without a hollow portion. The advantage of solid tires is that they do not puncture. However, they do not absorb the vibrations of the wheelchair, and thus they are better for the sort of wheelchairs which are used generally on flat surfaces, such as in public and commercial facilities.

The most common type of wheel is a wire-spoke wheel, which is similar to a bicycle wheel, and some wheels are made of carbon materials for weight-saving. Design also varies; some are very sporty and some are even made with illustrations. Figure 3-7 shows a wheel covered with a hub cap in order to prevent the fingers from getting caught in the spokes.

Figure 3-7 A wheel with an illustrated hub cap for children (OX Engineering)

Casters are usually used as front tires, and they are usually located near the foot support. The solid-rubber tire is a basic type of caster, while some are semi- or full pneumatic tires with a large and wide diameter, with suspension which provides shock absorption, quietness and a high-ground covering ability. However, tires made with a large diameter have a negative impact on the turning performance of a wheelchair.

Seat and backrests

The sitting posture on a wheelchair causes pressure concentration on the buttocks of the occupant, and this increases the chances of pressure sores developing. Thus, a seat with a high performance in relieving pressure is ideal. A seat with a good performance in this area can help the occupant maintain a good posture and high stability. For example, when using a manual wheelchair, the occupant requires back and forth movements of the upper body in addition to the motion of rolling the drive wheels with his/her upper limbs. Therefore, it is important that the buttocks of the user are properly supported. Basic seats are covered with cloth and secured to the frame of wheelchairs by rivets or screws, while some are attached with hooks and loop fasteners to increase removability. The cushion on the seat must be ideal for a long-distance ride. A relatively firm seat can help in reducing the fatigue of the user.

Figure 3-8 shows air cushions which can provide a flotation effect, and their pressure dispersion effect can prevent pressure sores from developing. Figure 3-8 (a) is a cushion made of 50 large and small air blocks. Medical silicon-blocks at the top layer of the cushion moderate the increase of body temperature of the occupant, and they absorb vibrations and shocks to the wheelchair while reducing the fatigue of the occupant. The slits in every direction suppress the

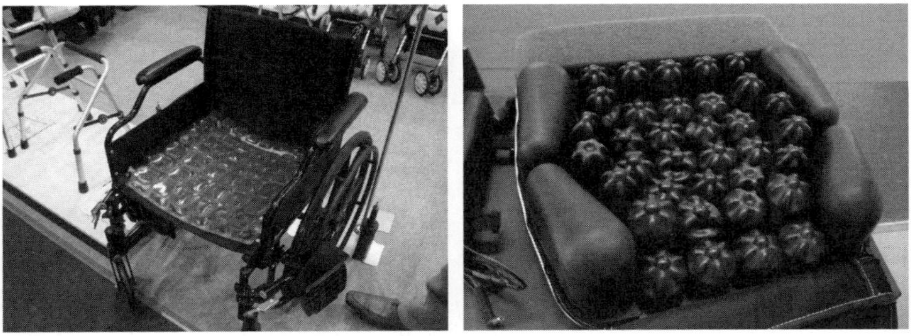

(a) Yuba Industrial　　(b) Yokohama Rubber

Figure 3-8 Cushions on wheelchairs

transverse repelling force, and this in turn will reduce the risk of developing decubitus. Moreover, the adjustability of air capacity in the cushion helps the occupant maintain a stable sitting posture. The cushion shown in Figure 3-8 (b) has an automatic computer control of air capacity, and this helps adjustment of any list in the body trunk of the occupant.

Backrests have two types. One is a fixed-type, of which the angle between the backrest and seat is not adjustable. The other is a reclining-type, and it is possible to tilt it in a backwards direction. A cloth sheet is often used as a backrest and it is normally fixed to the frame of a wheelchair. The thickness and the material of the cloth can usually be arranged according to the progress of the occupant's symptoms and body disfigurement. The height of the backrest can affect the user's movements when he/she rolls the hand rims of a manual wheelchair, and thus the highest point of the backrest is often set at around the lowest part of the user's shoulder blades.

In order to ensure the user's ability to move the upper body and both arms, some backrests are made short and they only support the hips.

Armrests and footrests

Figure 3-9 shows an armrest on which the occupant can rest his/her elbow and can still keep the balance of a sitting posture. Armrests come in fixed-, removable- and double-swing-styles. The double-swing-style is useful when the occupant transfers from a wheelchair to a bed, and the height of the armrest is best

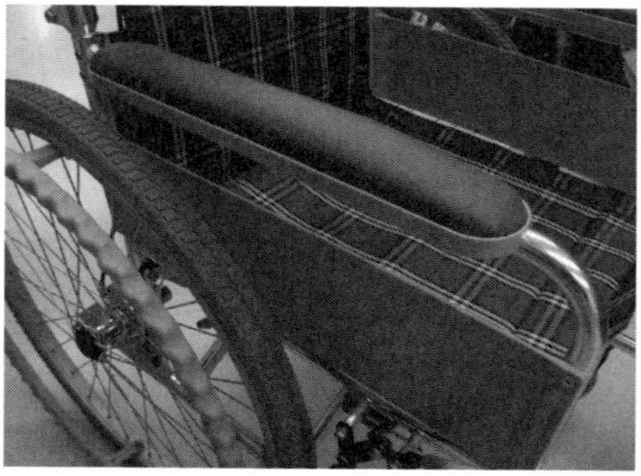

Figure 3-9 A fixed-style armrest

when the occupant can bend his/her elbow slightly. Armrests are also important for reducing pressure on the buttocks and thighs.

If the occupant needs to use his/her elbows constantly to support their body, choosing the right height and material of the armrests can also help in reducing pain on the elbows. According to some reports, 80 % of wheelchair users experience discomfort with their armrests (17). Inadequate armrests may cause, for example, a hunched position, excess elevation of the scapular arch, and pains and muscle fatigue. Basic armrests are fixed to the panels or the frame of the backrest of a wheelchair, while flip-up or other mechanisms similar to footrests are adapted to armrests in order to afford the occupant better entry to and exit from the wheelchair.

Footrests hold the foot of the occupant of a wheelchair, and usually the footrests are designed to be pivoted away from the centre line and folded up in order to help the occupant with entry to and exit from the wheelchair (Figure 3-10). If the soles of the feet are not placed properly on footrests, the position of lower limbs or lower thighs becomes unstable or the toes may be caught between the footrests. These situations can cause injury to the occupant.

The incorrect position of a footrest can cause leg malformation or other secondary disabilities to the occupant. Furthermore, the distance between the seat and footrests is a key factor in whether a comfortable sitting position is possible for the occupant. When the distance is set so that the back of the thighs lightly touches the surface of the seat, the body weight of the occupant can be well

Figure 3-10 Footrests, with one footrest pivoted upward

supported by his/her buttocks, thighs and soles.

Brake and grips

A brake-system is installed on basic wheelchairs, and usually a brake is equipped on each drive wheel of a wheelchair. There are basically two brake-systems for wheelchairs: the lever brake-system (see Figure 3-11) and the link brake-system. The lever brake-system is simpler mechanically, but it needs more power for braking a wheelchair. Generally, the brake mechanism becomes effective when pressure is put on the drive wheels directly. When a wheelchair is equipped with pneumatic tires, the amount of air inside the tires can have an impact on the efficacy of the brakes. In addition, other factors such as road conditions or the timing of putting on the brake can have an impact on brake efficacy.

Figure 3-11 A lever brake-system

Figure 3-12 A grip with a brake

Grips are installed behind the backrest, and these are generally used by wheelchair helpers. The height, location and strength of the grips are important, as grips are used when helpers provide special assistance to wheelchair users, for example, on slopes, uneven surfaces and stairs. The correct grips can prevent the physical burden of helpers from being unnecessarily increased. As shown in Figure 3-12, a brake is installed together with grips on some wheelchairs.

Frame and tipping lever

The frame of a wheelchair is mainly made of steel, aluminium, titanium or stainless steel. A frame made of steel is the heaviest among the frames made of these materials, and a basic wheelchair made of steel weighs 17-20 kg, while an aluminium-made wheelchair weighs 12-14 kg. A basic wheelchair made of titanium weighs around 10 kg. Steel is generally cheap and strong, and thus steel-made wheelchairs are widely used in hospitals and facilities for elderly people. High-priced wheelchairs are often made of titanium, which is light and strong. Carbon is also used for high-priced wheelchairs, since carbon is light and it has a high vibration-absorbency quality as well. In recent years, low-priced wheelchairs made of steel and aluminium have been imported from Korea, Taiwan, China and other Asian countries into Japan.

A tipping lever is used when wheelchair helpers push a wheelchair to mount uneven surfaces. When the lever shown in Figure 3-13 is pressed by the foot (which is used to apply the principle of leverage), the casters of the wheel-

Figure 3-13 A tipping lever

chair are lifted high enough to mount uneven surfaces. Some wheelchairs are built with a shorter lever, or with no tipping lever, in order to improve the design.

Engström summarised the dangerous parts of a wheelchair and the areas and types of injuries which may be suffered by the occupants of wheelchairs, and this is shown in Table 3-2.

Table 3-2 The dangerous places and the types of injuries

Dangerous parts	Area of injury	How injury occurs
Armrest plate	Thumb	A thumb is caught between a tire and an armrest plate.
Dust guard	Thumb	A thumb is caught at the sharp corner between a tire and a dust guard.
Brake	Thumb	A thumb gets caught between a tire and the brake.
Handrim and wheel	Thumb	A thumb gets caught while propelling the wheels.
Handrim hanger	Finger	Fingers get injured by the thin and sharp hand rim hanger when the user is slowing the speed of wheelchair.
Spoke	Finger	A finger gets caught between spokes and the wheel. A spoke guard can prevent the injury.
Tire	Thumb	A thumb gets injured by touching a tire.

Electric wheelchairs and Japanese Industrial Standards

Electric wheelchairs in Japan are mainly divided into two groups: a joystick-type (a self-operating standard-type) and a mobility scooter-type (a self-steering handle-type). The users of electric wheelchairs with joysticks are often those people who have impairments with their lower and upper limbs, and these impairments prevent them from operating a manual wheelchair. Many of these people suffer from injury of the spinal cord, muscular dystrophy or other generalized disabilities. Transport is essential for persons with disabilities when they participate in society, and thus electric wheelchairs play an extremely important role among persons with severe physical disabilities. For people with congenital disabilities especially, electric wheelchairs are usually an important means of active transport.

The users of electric wheelchairs with joysticks operate the wheelchair by controlling the joystick (see Figure 3-14). Similarly to manual wheelchairs, the chair rolls in accordance with the difference of rotation frequency among the two drive wheels which are installed at both sides of the chair. Some electric wheelchairs with joysticks have a power-steering function which is powered by an electric motor. This function permits the users to make turns when they control the directions of movement of the front casters directly. Moreover, a middle-wheel drive system has been newly developed and it has improved the turning performance of the wheelchair. The drive wheels are located right under the occupant in this system, and the chair rotates around the occupant when he/she tries to change direction. This creates a smooth and natural turning of the wheelchair. Incidentally, according to the Japanese Road Traffic Law, the speed limit for wheelchairs is 6 km/h or less, and basically the same conditions as for pedestrians are applied to the users of wheelchairs.

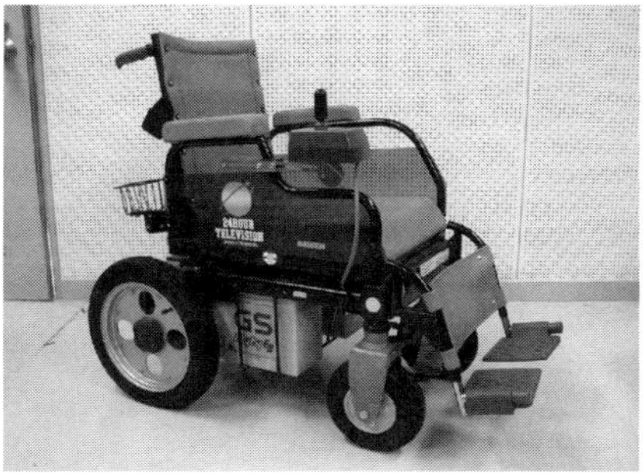

Figure 3-14 An electric wheelchair with a joystick (Imasen Engineering)

A mobility scooter is a single-passenger electric vehicle with three or four wheels (see Figure 3-15). The mobility scooter was originally developed for elderly people, and thus it is often called a "senior car." Driving a mobility scooter requires no driving license and operation is relatively easy. Therefore, it has become very popular among elderly people who experience difficulties in

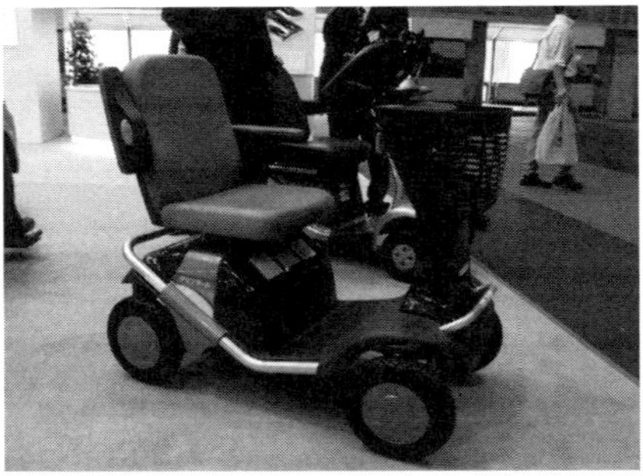

Figure 3-15 A mobility scooter (Suzuki Motor)

walking. According to the Road Traffic Act in Japan, mobility scooters are also regarded as equal to pedestrians. Therefore, mobility scooters must be driven on sidewalks.

Japanese Industrial Standards (JIS) is one of the collections of national standards in Japan. JIS was enacted based on Industrial Standardization (established in Japan in 1949) with the purpose of promoting Japanese industrial standardisation. Regarding wheelchairs, JIS specifies the degree of strength and operation performances required when wheelchairs are driven and parked. In May of 2008, the JIS Certification Bodies Association was established in order to examine the quality of products as one way to deal with an ageing society, and also to ensure the safety and comfort of people generally. Since then, manufacturers of wheelchairs can put a JIS mark on both manual and electric wheelchairs after their products have been certified by the JIS Certification Bodies Association (JISCBA). Usually, the JIS mark on wheelchairs is coupled with a mark (see Fig-

Figure 3-16 JIS mark with additional welfare tool mark

Figure 3-17 TS mark

ure 3-16) which indicates that it is a welfare tool. The same system is applied to other products, including home-care beds, and products with the welfare mark are exempt from consumption tax. In 1992, the TS (traffic safety) mark (see Figure 3-17) was also introduced in Japan. The mark is attached to products which have satisfied the standards specified by the Road Traffic Act and have been certified by the National Public Safety Commission (of Japan) for its model code.

Sidewalks

Basic configuration

As a general rule, sidewalks should afford safe and smooth transport circumstances to all pedestrians, including wheelchair users, elderly citizens and persons with visual impairments. The following are the basic rules used when sidewalks are constructed in Japan.

(1) The valid width of a sidewalk should be 2 m or more, and the width should be 3 m or more for a sidewalk designed to be used by both bicycles and pedestrians. The width has to be the same all the way from the beginning to the end of the sidewalk.

(2) In principle, the sidewalks have to be permeable.

(3) Sidewalks are, in principle, built with 5 % or less longitudinal gradient, and 2 % or less transverse gradient (it is 1 % or less in the case of permeable sidewalks).

(4) Each sidewalk is to be marked off by curb-stone in order to secure the safety of persons with visual impairments.

(5) The height of the curb-stone must be 15 cm or more, and any planting zone, line of trees or fence should be set up on the boundary between the street and sidewalk if it is required.
(6) The surface of a sidewalk should be 5 cm off the ground generally. However, the height is to be determined by taking account of the neighbouring circumstances and conditions, such as adjoining houses which use certain portions of sidewalks when cars drive into or out of the dwellings.
(7) When sidewalks are connected to a pedestrian crossing, the face of the sidewalks should in principle be 2 cm off the grounc.

Maintenance and improvement of sidewalks and super-elevation

Flattening of the surfaces of sidewalks is a key concern when any sidewalk is improved, so that it becomes a barrier-free zone. Currently, when sidewalks are constructed, slopes are purposefully built on some sections of sidewalks in Japan. This is to help the entry and exit of cars belonging to adjoining houses, which drive over a certain section of a sidewalk every day. This circumstance has created the "rolling" phenomenon of sidewalks, where different heights and gradients on the surface of sidewalks occur almost alternatively. Sidewalks under such conditions are one of the many serious risk factors for wheelchair users, as this prevents them from stable driving. Improvement should urgently be made to sidewalks in such a condition. Furthermore, another important task for improving and maintaining road conditions is the reduction of the difference in

Figure 3-18 Rolling of sidewalks

the level of surfaces between pedestrian crossings and streets. The difference in level between pedestrian crossings and streets should be minimised, and the safe as well as stable usability of sidewalks and streets should be created, as it is a basic concern when a barrier-free environment is created or improved. Barrier-free circumstances which are created and improved due to such considerations may afford stable travel to wheelchair users, and this may reduce their physical burden.

However, if sidewalks are improved by eliminating the rolling phenomenon but a transverse gradient remains as a countermeasure for rainwater drainage, then the condition of the sidewalks still remains in a state where the physical burden on wheelchair users cannot be reduced. According to a study conducted by the author, a transverse gradient causes wheelchairs to move to one side, and consequently the occupant of the wheelchair must constantly make an effort to keep the wheelchair straight. This effort increases the physical burden on the wheelchair occupant. The burden increases significantly when a wheelchair is going downhill on a transversely-graded sidewalk. This happens because the longitudinal gradient and the transverse gradient increase the physical burden on the occupant, and the burden then affects his/her heart rate. Even if the burden is on only one part of the body, it still affects the physical condition of the occupant while he/she is operating a wheelchair. Traffic facilities are accessed and used by people with various kinds of impairments. Therefore, when decisions are made on the creation of well-considered and well-balanced traffic facilities, those decisions should be made in accordance with a good understanding of the evaluations and examinations conducted on the behavioural characteristics of persons with impairments.

Barrier-free access and evaluation approaches

Today's barrier-free programs are implemented for the development of public facilities and other circumjacent facilities. However, the programs are developed as one of the goals for improving the entire range of traffic facilities. Unfortunately, planning methods needed for creating barrier-free environments are still swayed by financial limitations, orders of priority and planning stages, and even the methods employed for conducting evaluations remain unclear when it comes to a barrier-free environment.

When evaluations are made of barrier-free facilities, for example, those evaluations must be made of the facilities themselves, along with technology, economy, amenity and other issues. Then, a comprehensive evaluation based on these categories can be made and reported. In addition, the cost benefit is evaluated and survey researches are conducted and studied. However, very few evaluations are made based on surveys conducted among the users of barrier-free facilities. Not only pursuing effectiveness and efficiency, but how to interface between vehicles and people and how to evaluate the environment around people are also matters of concern when road infra-structure is to be improved. When barrier-free circumstances are created, the characteristics of persons with impairments, their wheelchairs and other devices for assistance have to be taken into consideration. Furthermore, efforts have to be made on reducing the physical burdens of those people who make access to and use such facilities, and evaluations and judgements have to be made concerning the degree of psychological comfort which is felt by the users.

Uneven surfaces and gradients

Generally, certain differences of level are created intentionally on the borderline between sidewalks and streets. Such differences help, for example, persons with visual impairments to distinguish the sidewalk from the street. In present-day Japan, the standard difference in level between a sidewalk and a street is set as 2 cm, according to the "Guidelines for maintenance and improvements of the easier accessibility and usability of transportation infrastructure" which has been set up, based on the Barrier-free Transportation Act [18]. However, the structure of the outside edge of sidewalks and streets is to be studied and determined. This is to be based on the evaluations and results concluded from efforts to understand the behavioural characteristics of people who include wheelchair users, persons with visual impairment and elderly people, as well as of the architecture of sidewalk edges of streets.

Differences in level are found not only on sidewalks. The differences are found basically everywhere inside and outside of buildings, and some improvements have been made to create circumstances in which wheelchair users feel less inconvenience. However, such progress is still being made at a snail's pace. For example, people sometimes find the entrances of buildings are designed with stairs that are high off the ground, even though the inside of the building is designed with the concept of barrier-free access.

In inner-city areas, safe and smooth transportation by, for example, wheelchairs, has become easier to a certain extent. However, there are still very many problems in suburbs and on residential streets. Topographically, there is less flat land and more pitched and uneven areas in Japan than average, and there are many areas which wheelchair users cannot drive through without the help of other people. Building ramps are one of the common ways to eliminate a difference in level on roads. However, creating ramps requires a big piece of land on each occasion. For instance, a road surface which is 1 m off the ground will need at least 7-12 m length on the bottom surface for the creation of a ramp. If there is not adequate space for building a ramp, a stair lift, such as the type shown in Figure 3-19, is an effective way to eliminate differences in levels. Stair lifts can be used as stairs, and when wheelchairs drive onto them, they can be transformed from stairs to a flat surface. It is possible to make the handrails and stair treads of wood for home use.

Figure 3-19 A stair lift (Koyo Engineering)

Road surfaces of sidewalks

Sidewalks require regular maintenance in order to prevent their slow deterioration. For example, if a sidewalk is provided with no regular maintenance,

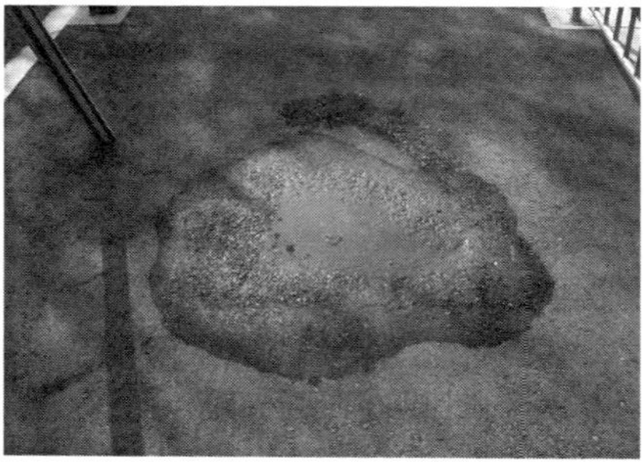

Figure 3-20 A puddle of water made in the depression on a sidewalk

Figure 3-21 The concave-convex surface of interlocking blocks

raindrops will collect in the concave-convex surface of the sidewalk (see Figure 3-20), and this becomes an obstacle for wheelchair users and other persons with difficulties. Moreover, on sidewalks which are coated with interlocking blocks, if or when the blocks become depleted, the joints create depressions and this creates difficulties, especially for wheelchair users and persons with visual impairment (see Figure 3-21).

Sidewalks are paved according to various methods and three representative examples are as follows:

(1) Asphalt pavement

An asphalt mixture is the most common material for pavements in Japan. The mixture is used for both streets and sidewalks. In addition, an asphalt mixture is sometimes used as a permeable pavement. An asphalt pavement is resilient, and when the pavement requires repairs, only a short time is required before the pavement is placed in service again. However, if one of the advantages of asphalt pavements is that they are easy to repair, then the disadvantage is the fact that they are easily deformed under high-temperature conditions, such as in summer.

(2) Concrete pavement

Cement-concrete is the main material for concrete pavements and it is used for both streets and sidewalks. A concrete pavement is hard and not easy to break. It also has the quality of lasting for a long time, as well as good heat tolerance. However, compared to asphalt pavements, concrete pavements are easier to break and a longer time is required for repairs.

(3) Block pavement

A block pavement uses a variety of materials, and local repairs are easy to do, while excellent colours and designs are available. Construction of a block pavement, however, requires a long period of time and a greater labour supply. Furthermore, such a pavement is weak under a heavy concentrated load, and such loads often cause mounding and concaving of the blocks.

Shown in Figure 3-22 is the grating cover of a drain ditch. The conventional grating covers in Japan have grid-like holes, and each hole has a size of 1 cm or more for a better drainage performance. However, the size of the holes were big

Figure 3-22 Clogged grating holes with fallen leaves

enough to create problems for people who use wheelchairs or walking sticks, as the holes catch the bottom part of a wheel or the end of a stick. Consequently, to prevent wheels from falling into the holes, the grating covers have been replaced with covers with smaller grid-like holes, the size of which has become less than 1 cm. However, stones, fallen leaves and dirt collect easily on the holes and this has created difficulties in the maintenance of grating covers.

Footsteps and slopes

Many station buildings are installed with elevators in present-day Japan, and most of the elevators are designed to help people be brought up and down directly between the ground floor and the ticket barrier. Shown in Figure 3-23 (a) is a wheelchair climber. This chair basically can ascend or descend any stairs. A sensor system is installed to the wheelchair climber, and the sensor detects and checks lean and other conditions required for ascending or descending, for example, stairs. Shown in Figure 3-23 (b) is an electrically-controlled wheelchair climber. Two wheels and two robot-arms are activated in alternate shifts, and the wheelchair ascends and descends stairs similar to a person walking. The drive section is unified with the wheelchair, and it saves the trouble of removing or installing a drive section.

Building ramps is one of way to eliminate differences in level on the ground.

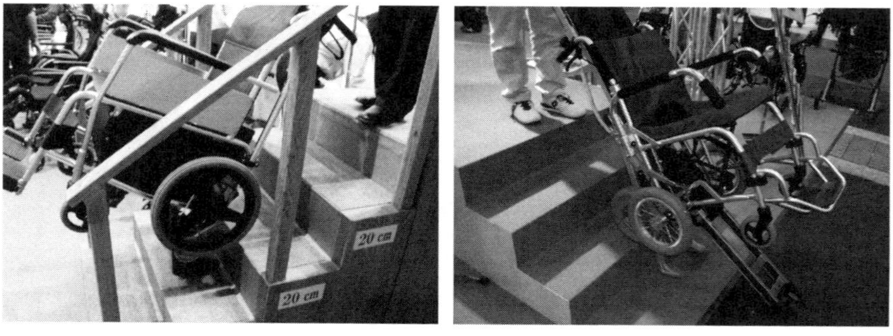

(a) TS Technology (b) Nabtesco

Figure 3-23 Wheelchair climbers

Figure 3-24 Flat area on a ramp

In Japan, the gradient ratio of "ramps which are an alternative to stairs" is stipulated as 12 % or less in Article 25 of the Order for Enforcement of the Building Standards Act. However, this degree can cause physical difficulties for the users of manual wheelchairs. The gradient ratios of ramps are to be 8 % or less for interiors and 5 % or less for exteriors, according to the New Barrier-free Act. However, in cases where the difference in level is 50 cm or less, the gradient ratio can be reduced. The centre of gravity of a wheelchair is high, and thus there is the danger of a wheelchair falling backwards if the gradient of ramps is sharp. The ideal gradient ratio is 5 % or less. Even with that gradient ratio, the occupants of wheelchairs will find that they need to have strong arms and a large capacity for endurance as the ramps get longer, while ordinary pedestrians find them relatively gradual. If a landing is not built on a ramp and if the occupants of wheelchair stop moving because they are tired, it is quite dangerous as the wheelchair reverses unless the brake is put on properly. As shown in Figure 3-24, landings are needed to be built so that the occupant can take an occasional rest.

The bigger the gradient ratio becomes, the stronger the muscle activity which is required when the occupant needs to grab or hold the hand rims of the wheels. If the gripping force on the hand rims is weak, wheelchairs tend to move downwards on upward ramps. In such a case as Figure 3-25, the occupant has to operate the wheelchair with the head bent forward constantly, and this makes the occupant feel more fatigue in the shoulders, arms and

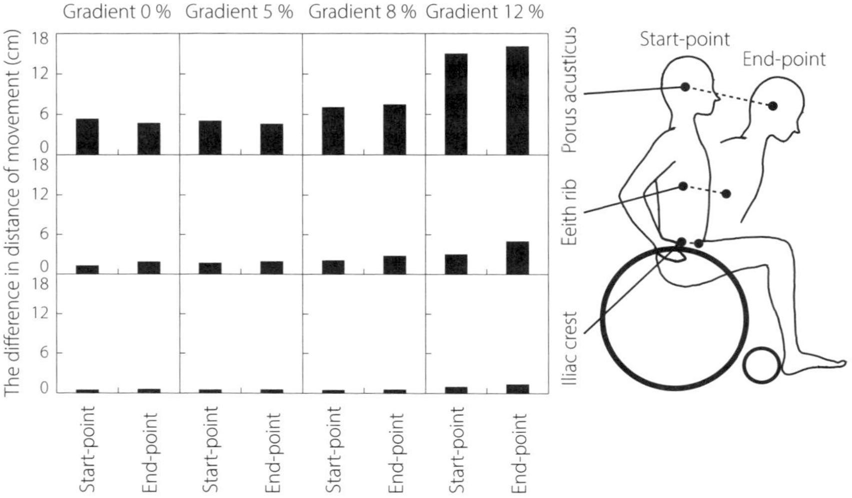

Figure 3-25 The difference in distance of movement of parts of the body in various gradient ratios on ramps

fingers (19). When the occupant is in such a bent posture, he/she will have difficulty of reaching out to the hand rims, and this causes his/her elbows and upper arms to move upwards and sideways. This unnatural posture causes fatigue in his/her shoulders, arms, fingers and other parts of the body.

CHAPTER FOUR

Wheelchair users and cars

The environment for driving cars

The acquisition of a driver's license by persons with physical disabilities was legalised in 1960 under Article 88 of the Road Traffic Law, and this allows persons with impairments to their upper limbs to obtain a driver's license. In the case where the driver of a car is a person with severe disabilities in their four limbs or body trunk, which creates disadvantages in driving, there is a condition requiring the installation of driver-assist devices in accordance with the degree of impairment [(20)]. In November 2000, the Barrier-free Transportation Act was implemented as an act for public transportation accessibility and usability by the elderly and physically disabled persons. In modern Japan, as a social issue, great attention has been drawn to plans and practices for making the roads and facilities barrier-free zones as part of an effort to support the independence of persons with physical disabilities. According to a report proposed by the Japanese police in 2011 [(21)], 251,686 persons with physical disabilities were granted a driver's licence with some conditions. Of this total, 39,683 people were supplied with a licence under the condition of driving with an acoustic aid, 4,089 under the condition of driving with an artificial arm or leg and 207,416 under the condition of driving cars modified specially for persons with physical disabilities.

Almost all wheelchair users find that cars are a highly-convenient transportation means in comparison to trains, buses and other public forms of the transportation system. Cars can usually provide these people with "door-to-door" transportation. The drivers of cars do not need to consider or worry about using steps and sloping roads, or facing certain obstacles which they often encounter when they are travelling by wheelchair. They are more likely to use cars than public transportation if it is possible. Therefore, cars are an important transportation means for persons with physical disabilities in order for them to participate in society independently. However, as J. F. Dols et al. [22] and P. Kember [23] have pointed out, the system of operation and environment for using cars do not allow for the physical impairments of persons with disabilities, even if their cars are remodelled especially for them. The transportation environment of roads is usually designed based on the assumption that they will be used by healthy people, and almost no consideration is given to ensuring an ergonomically safe and comfortable transportation environment for the disabled. For instance, the provision of road signs or other information is generally targeted at healthy people, and the main purpose of the signs and similar information is to help healthy drivers to drive safely and minimise risk. In the current situation, there is no consideration of safety and risk management developed in accordance with the needs of persons with physical disabilities in Japan.

In America, a membership-based club, the American Automobile Association (AAA), was established in 1902. Since then, the AAA has made all-round efforts in securing road safety and they have also published the Handicapped Driver's Mobility Guide. In addition, the AAA provides a lot of support which is beneficial to persons with physical disabilities. The support provided by the AAA include the tasks of selecting cars and factories, installing driving-assist devices and their maintenance, providing education and introducing insurance policies necessary for driving cars. The main companies and organisations which are involved with AAA activities are: major auto companies, assist-device manufacturers, driving schools, the United States Department of Veterans Affairs, the Society of Automotive Engineers, the Rehabilitation Engineering Association, the Association of Driving Educators for the Disabled, and the American Driver and Traffic Safety Education Association. A large number of activities are conducted literally by joint initiatives between public and private sectors.

Acquisition of a driver's license by persons with physical disabilities

In Japan, attending a driving school is compulsory for the acquisition of a driver's license, and the same system applies to persons with physical disabilities. However, persons with physical disabilities are required to have a preliminary interview with personnel who conduct an official consultation and audition. The consultation and audition are conducted at the Driver's Licence Centre and the Driver and Vehicle Licensing Centre. Persons with physical disabilities are expected to take an aptitude test to determine whether they have reasons for disqualification, as stipulated by Article 88 of the Road Traffic Law. Disqualification is applied to persons with the illnesses described below.

(1) Persons who suffer psychological illness, intellectual disabilities, epilepsy, a severe visual impairment or are orally-challenged.
(2) Persons with physical disabilities which are specified by the Cabinet Order (Article 33).
    - Persons who have lost upper limbs above the elbow joints, or persons who have no upper limbs or no functional upper limbs.
    - Persons who have impairments in the lower limbs or body trunk, and who are incapable of sitting.
    - In addition to the above, persons who can not operate steering or other devices at will.

Even if the disabled qualify, they will be under conditions which specify license class, type of vehicles and assist-devices to be used.

In addition, in the case of those persons who have become disabled after they obtained a driver's licence, they are then required to take an aptitude test. The results of the test for such people are categorised as follows:

(1) Unconditionally qualified: they are permitted to drive as they did before.
(2) Conditionally qualified: they are provided with a driver's license with certain conditions. (They are permitted to drive a car which satisfies these certain conditions.)

(3) Ineligible: they are required to take an aptitude test after completing rehabilitation programs.

Figure 4-1 shows the process of obtaining a driver's license for persons with physical disabilities in Japan.

For example, the conditions for a license given to persons with impairments in the lower limbs are described below.

(1) For persons who have lost lower limbs below the hip joint, or persons who have no function in the lower limbs.

- Types of license given

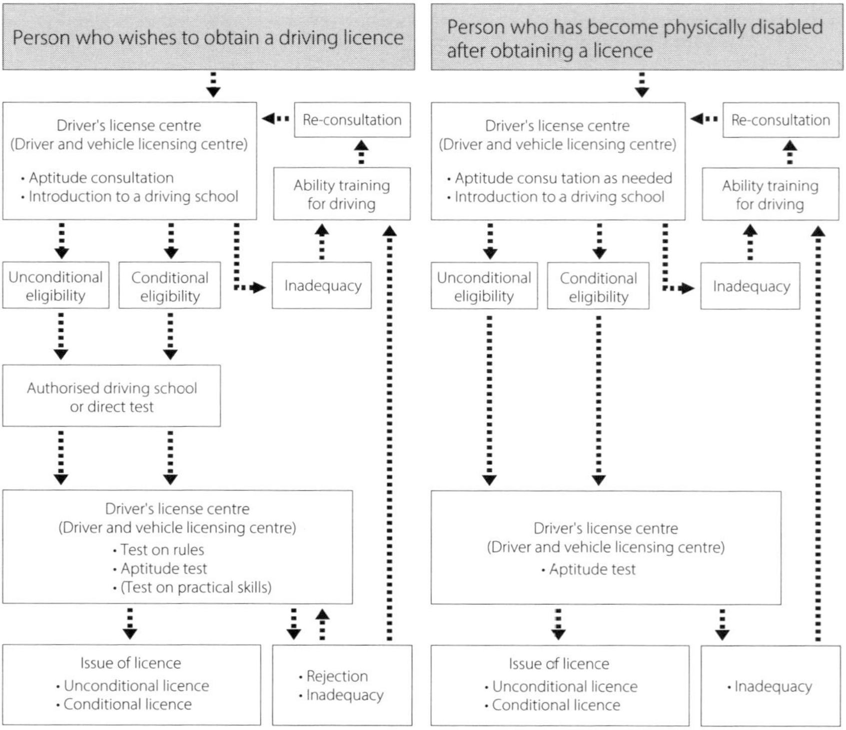

Figure 4-1 Process of obtaining a driver's license for the disabled in Japan

Basic license, small-sized special motor-vehicles, motorcycles with a motor engine of 50 cc or less

- Conditions for obtaining the license

  Types: Hand-driven cars (a model in which acceleration and braking can be operated with the upper limbs), automatic cars, small motor-vehicles with three or four wheels

  Physical condition: Artificial legs are compulsory, and the artificial legs must be effectual for driving.

(2) For persons who have lost lower limbs below the knee joint, or persons who have significant impairments to their function.

- Types of license given

  All types (except large motorcycles)

- Conditions for obtaining the license

  Types: Limited to automatic cars, hand-driven automatic cars and small motorcycles depending on the users' physical and technical abilities

  Physical condition: Artificial legs are compulsory, and tools to compensate for the function of lower limbs are necessary.

Special marks for persons with physical disabilities

A special mark for persons with physical disabilities has been introduced since a review was made of the reasons for disqualification of persons with disabilities from obtaining a driver's license in 2001. The mark for persons with physical disabilities (also known as the Clover Mark in Japan), which is shown in Figure 4-2, is not obliged to be attached to vehicles which are driven by persons with

Figure 4-2 Mark for persons with physical disabilities

disabilities. However, when there are risks that their impairments may cause difficulties in their driving, the person with the disability is expected to attach two marks, one at the rear and the other at front of the car, in places with a high visibility. Moreover, any cars which cause trouble to cars with the Clover Mark can be fined.

Shown in Figure 4-3 is the International Symbol of Access which indicates that the buildings and facilities displaying the mark offer easier accessibility and usability to persons with disabilities. The rules and usage of the mark are defined in the guidelines issued by Rehabilitation International. The subjects of this mark are all persons with disabilities, and thus it is not limited to wheelchair users.

Figure 4-3 The International Symbol of Access for persons with disabilities

Parking spots for the disabled and the system to exempt the disabled from parking prohibition rules

Those persons with disabilities who use wheelchairs need to transfer their wheelchair to and from their car, and this action requires a certain amount of space. As shown in Figure 4-4, parking spots for the disabled are usually made spacious so that transferring a wheelchair is possible with the car's door opened. In present-day Japan, more and more parking spots for the disabled are available, and these spots are indicated with a symbol. However, healthy people sometimes park their cars on the spots and this has become a serious problem. Thus, it is important to advocate for the awareness of the symbol, to help the general public understand the meaning of the parking spots with the mark.

Figure 4-4 A parking spot for the disabled

In addition, there is a system to exempt the disabled from some, but not all, parking prohibition rules. This system was introduced to help persons with disabilities park their cars at the point nearest to their destination when it is necessary. Permission can be granted by placing a card inside the front of the car which indicates that the vehicle is exempted from parking prohibition rules (see Figure 4-5). The card is issued by the National Public Safety Commission. This system is applied to people who have registered at governmental organisations as persons with disabilities. These people are supplied with booklets or certificates to prove that they have a certain degree of disability.

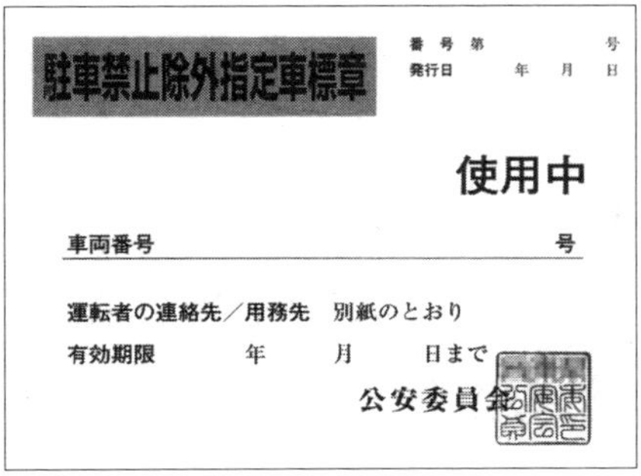

Figure 4-5 Card which indicates that the vehicle is exempted from parking prohibition rules

Subsidy programs

Cars belonging to persons with disabilities are sometimes exempted from automobile-acquisition tax, auto tax and light-vehicle tax in Japan. If or when the owner of the car is, for example, a family member, documents are required to be submitted to municipal offices to prove that the disabled person is living with the vehicle's owner and that the car is used mainly for going to see doctors.

A subsidy is given when a car is remodelled specially for persons with disabilities. However, the amount of the subsidy depends on the degree of disability, and the person needs to apply for the subsidy at the municipal welfare office in advance. For example, approximately 100,000 Japanese yen is provided in cases of remodelling the steering wheel or the hand-operating device, though the amount varies depending on the municipality. The subjects of the subsidy are those persons with disabilities whose age is 18 or more and who have a first or second grade certificate of physical disability. Moreover, the car must belong to a person with a disability and must need remodelling for their use. The vehicle owner's income of the previous year also affects the amount of subsidy the person receives.

There is an additional system which offers discounts given to motorway fares and other toll fares. This discount system was introduced with the purpose of supporting persons with disabilities so that they can participate independently in social activities, including such activities as going to work, schools and doctors. On December 1, 2003, the discount system on motorways for persons with disabilities was amended, and they can use motorways at half price if they go through the necessary procedures at the municipal welfare office in advance

The subjects of the discount are those who have an identification booklet for the physically disabled, and the discount is also given to a driver who is driving with and for the person with the certificate.

There are other types of discount systems for persons with disabilities, including welfare-taxi discount tickets and gasoline tickets. These systems are different depending on the municipality. These discount systems are available for those who can submit the necessary certificates and documents to the municipal welfare office.

Assist devices for driving which are tailored to each type of impairment and the selection of vehicles

Assist devices for driving, which are tailored to the level of a person's impairment and residual functional abilities, allow persons with disabilities to drive cars in almost the same way as unimpaired people. Auto-makers and companies which manufacture assist devices have developed devices for driving which are responsive to various types of impairments. The relation between the functions of the four limbs and assist devices for driving is summarised in Table 4-1.

When purchasing a car, an auto-shop which has a maintenance factory in-house is ideal, and it is better to choose a shop in a local neighbourhood. Moreover, selecting a vehicle requires careful consideration. For example, a car with a great deal of height has a good range of view, but it is inconvenient for transferring or storing a wheelchair. Sports cars, or similar models of low-ride cars, are highly manoeuvrable, but the transferring or storing of a wheelchair is difficult. Large-size cars have a big storage space, but they do not have a small turning circle. Newly-developed light cars tend to have wider amounts of space inside, and they have an ability to turn in a small radius. However, overall, the ideal type of car is a station wagon of which the engine size is around 2,000 cc and which is equipped with power-steering and power-window systems. Fleet cars have an advantage as a variety of components are available and easy to obtain. The choice of vehicles is greatly influenced by the conditions of impairment of the disabled. Therefore, it is ideal to seek advice from, for example, an occupational therapist, a physical therapist, experts in rehabilitation engineering and auto-manufacturers who have good experience of remodelling cars for the disabled.

Hand-operated assist-devices for driving

Fuji Auto Company was the first auto-company to develop a driving device which can be operated by hand in Japan. The device was first developed for the developer himself, who had had a car accident and suffered impairments to his hands. Since then, a variety of types of driving-assist devices have been developed in accordance with various types of impairments.

Table 4-1 Assist devices for driving tailored to each type of impairment

Assist device	Body site with impairments										
	Both upper limbs	Both lower limbs	Hands and fingers Both lower limbs	Right lower limb	Right upper limb Right lower limb	Right upper limb	Right upper limb Left lower limb	Left upper limb Right lower limb	Left upper limb Left lower limb	Left upper limb	Left lower limb
Hand-control system		●	●								
Steering grip		●	●		●	●	●	●	●	●	
Foot-operation steering	●										
Acceleration pedal on left side				●	●			●			
Direction indicator on left side			○		●	●	●				
Foot-operation direction indicator	●				○	○				○	
Foot-operation light-switch	●				○	○				○	
Light-switch on left side			○		○	○	○				
Foot-operation horn	●										
Lever for windscreen wiper on right side								●	●	●	
Parking brake on right side								○	○	○	
Parking brake which is operated by hands		○			○		○				○
Parking brake on right side	●										
Device for storing a wheelchair		○	○								
Transferring board		○	○								

Note: ● is compulsory, while ○ is something better to have.

For instance, persons with impairments to their lower limbs need to drive cars by using only their upper limbs, and the accelerator and brakes of the car are controlled by hand. In such a case, a device with a hand-control system, as in Figure 4-6, known as a floor-type, is commonly used. With this type, the device and a control lever are installed on the floor of cars with a clasp which plays as a fulcrum point on the lever. The lever is usually set at the left-side of the driver in Japan. Acceleration and braking are controlled by pulling the lever backwards and pushing the lever forwards, respectively. Some devices have switches for the direction indicator and head-lamps built in. The steering system is operated by the upper limb(s). The device has a function with an attitude-control system which operates during driving. However, if the driver suffers from motor paralysis or sensory paralysis, there is a risk of his/her lower limb on the device-side touching the device. Fixing lower limbs with a belt or using a cushion to lift the position of the lower limbs are required in such a case.

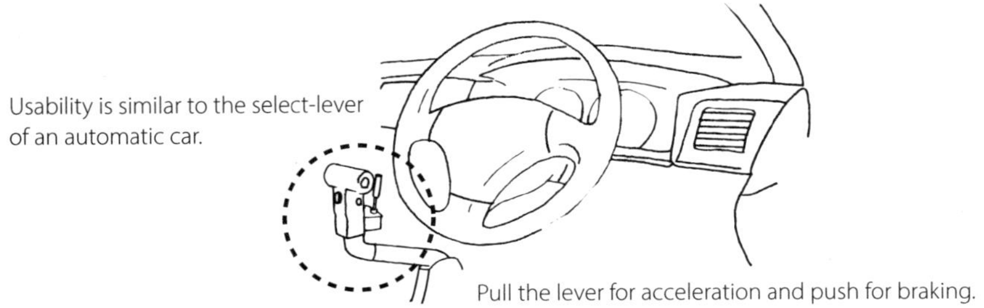

Figure 4-6 A device with a hand-control system known as a floor type

There is another type of device called a column type. This type of device has a hand-control system installed on the steering column, as shown in Figure 4-7. With this type, the device and a lever are installed on the steering column with a clasp which plays as a fulcrum point on the lever. This system enables the driver to have more space in the foot area of the car. However, a good distance has to be kept between the lower limbs and the steering or the clasp on the steering column in order to avoid contact. It is possible that such contact can cause injuries. The operating portion of the device is similar to that of motorcycles. If the drivers of such cars fitted with the device can move their fingers well enough to

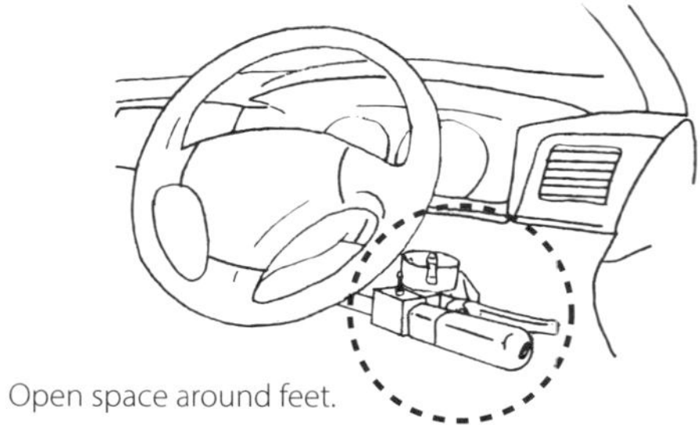

Figure 4-7 A hand-operated driving-device known as a column type

control it, then they can use a type of device which has an acceleration system built in. This type with an acceleration system has an acceleration ring installed inside of the steering wheel. While controlling the acceleration ring with the fingers, the wheel can be turned, and the control of acceleration is possible with either of the hands.

Recently, auto-makers have developed hand-operated driving-devices (see Figure 4-8). Models which are specially developed by auto-makers have the device installed inside of the centre console of the car, and this allows the drivers to have more space in the foot area than in those models in which the device is installed after the assembly of the car has already been completed. With the

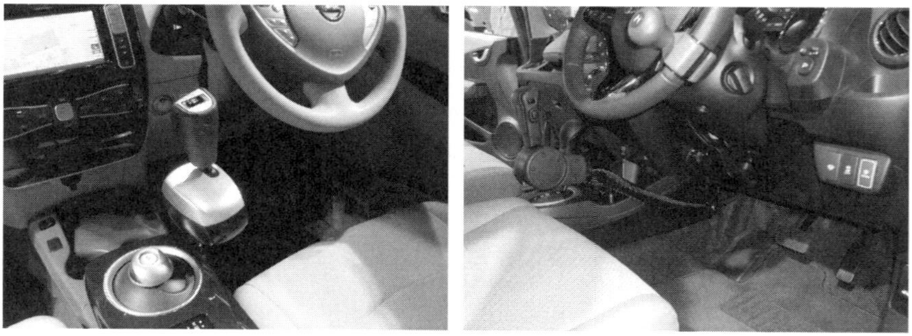

(a) Nissan Motor (b) Honda Motor

Figure 4-8 Hand-operated assist-devices for driving which have been developed by auto-makers traditional models, when the lever of the device is pushed or pulled, the entire lever moves mechanically. However, if the models are developed by auto-makers, a computerised control-system is installed and the system can be operated merely by moving a knob attached to the lever. This provides a stable driving posture even for persons with some impairment to their elbows.

Those cars which have the device shown in Figure 4-9 installed are called "joystick cars". They are models which have been developed for disabled people who cannot operate cars even with help from a hand-operated assist-device. The joystick model can be driven with one hand or one leg. The acceleration and braking systems are controlled by moving the lever forwards and backwards, while the wheel is turned by inclining the lever to the right or left. The shape and position of the controlling lever and the possible proliferation of switches are designed with all types of users taken into consideration. Furthermore, it is designed so that the user's hand will not slip off from the control platform while driving.

Figure 4-9 A joystick type assist-device for driving (Fuji Auto)

Driving seats

Impairments to the lower limbs or body trunk prevent drivers from keeping the body balanced and maintaining stable operation of the steering wheel and other driving systems in vehicles. When driving on winding roads in particular, drivers usually have difficulties in getting the car at the right speed and into the

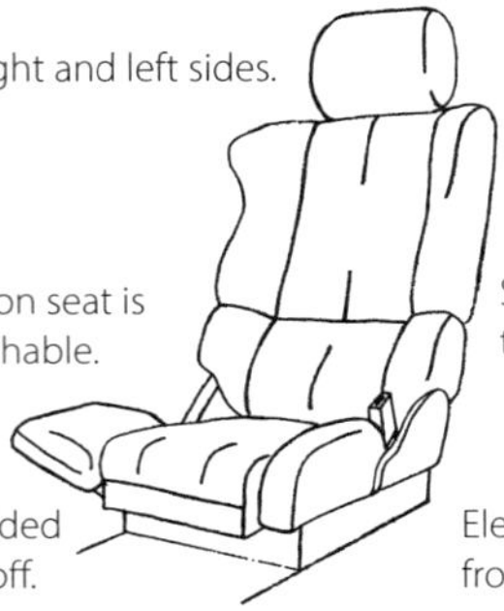

Figure 4-10 A driving seat for persons with physical disabilities

right travelling position, as Koji Endo has pointed out [24]. As a solution to such problems, special driving seats have been developed for persons with disabilities. An illustration of such a seat is shown in Figure 4-10. The functions of the seat aim to improve the physical conditions, safety and comfort of drivers who have impairments. The main functions are: (1) Improving the transferability of a wheelchair between the inside and outside of car, as well as the transferability of the driver between seats inside the car; (2) Improving the retention of the body trunk during driving; (3) Improving controllability; (4) Preventing the driver from developing decubitus; and (5) Providing incontinence control.

Figure 4-11 shows a body belt—not a seat belt—which is used when the driver has difficulty in maintaining the balance of his/her body trunk. The body

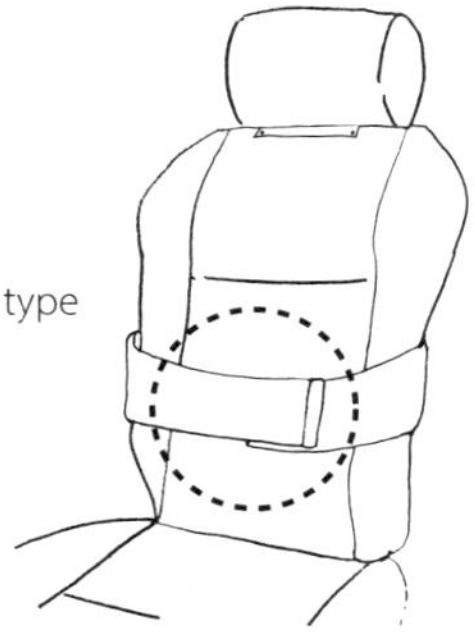

Figure 4-11 A body belt with a loop fastener

is looped with the belt and this is fixed with a loop fastener. This belt can help hold the body to the seat. However, leaning forward or other bending actions are not possible for the driver when the belt is on.

Turning devices

A turning device is an assist-device for persons with impairments to the upper limbs who cannot hold the steering wheel of a car. There are a variety of turning devices as shown in Figure 4-12, and drivers with impairments can choose one in accordance with the level of their impairment.

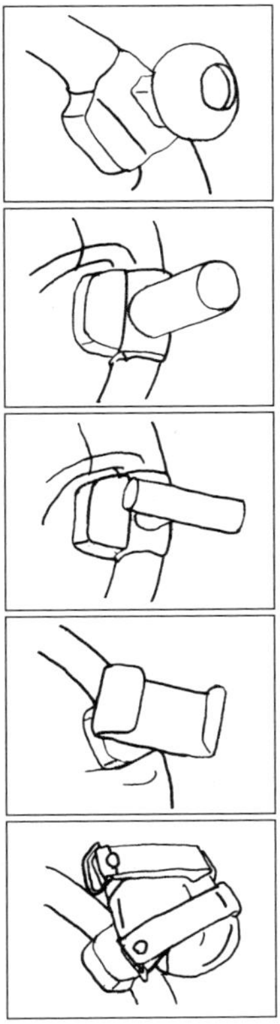

(1) Knob type grip:
Used by people without problems in the palm.
Similar to the grip of a fork-lift.

(2) Vertical-stick type grip:
Used by people with problems of abnormal palms or lack of grip power.

(3) Horizontal-stick type grip:
Used by people with problems of abnormal palms or lack of grip power.

(4) U-shape type grip:
Used by people among (2, 3) and (5), and also when there is difficulty to operate grip smoothly.

(5) Palm type grip:
Used by people with a lower level of grip power compared to (4).

Figure 4-12 Turning devices

Foot-operated assist-devices for driving

Persons with impairments to the upper limbs can drive a car by using an assist-device which is developed especially for them. The device helps such people operate the steering-wheel, accelerator, brake, direction-indicator and other operations required when driving. Figure 4-13 shows a pedal which helps operate the wheel by controlling it with the left foot. The device is usually installed on the left side of the foot space.

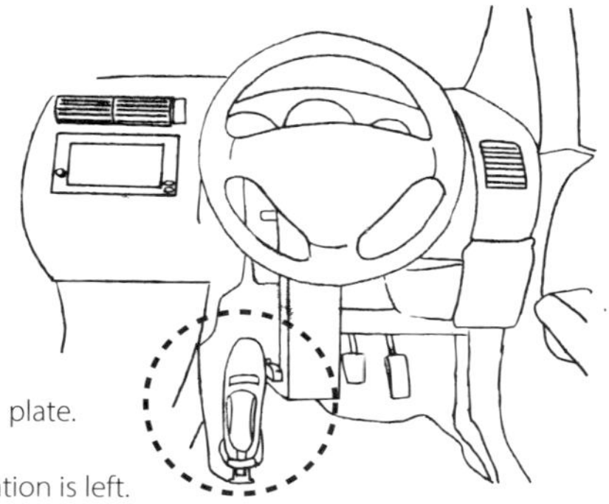

Figure 4-13 A foot-operated device

Figure 4-14 shows a device which is used by drivers who have impairments to a right lower-limb and thus cannot use the accelerator pedal, which in Japan is usually used with the driver's right foot. The pedal of the device is coordinated

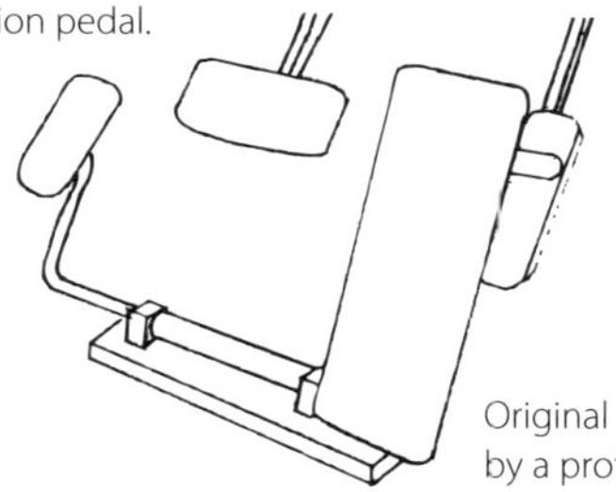

Figure 4-14 An accelerator-pedal for left foot

with the accelerator-pedal of the car. It is removable, and thus the accelerator-pedal can be operated in a basic way.

Figure 4-15 shows combination switches. Many switches used during driving are collected and located at the foot area. The switches are installed in the foot areas following a purchase order made by the users, and thus the switches are arranged in accordance with the physical conditions of the drivers. In addition, some cars are installed with a device which helps the driver fasten the seat-belt automatically, as is shown in Figure 4-16. Persons with impairments to their upper limbs have difficulty in putting on a seat-belt by themselves. The device is installed with a mechanism by which the seat belt is put on the driver automatically when the door on the driver's side is shut.

Figure 4-15 Combination switches for the foot

Figure 4-16 A device which helps the driver fasten the seat-belt automatically

Transferring

Transferring from a wheelchair to a car requires much effort from the users of wheelchairs. It is even more so in circumstances such as transferring on a busy street, as this requires the person to do so very quickly. Persons with impairments to their upper limbs have difficulty in transferring when lifting their bodies. These persons usually use a transfer-board for shifting between the car and wheelchair. The transfer-board shown in Figure 4-17 is a folding type. The board eliminates the gap and the difference in level between the driver's seat and the wheelchair, which is placed by the door. The driver slides on his/her buttocks to get onto either the driving seat or the chair.

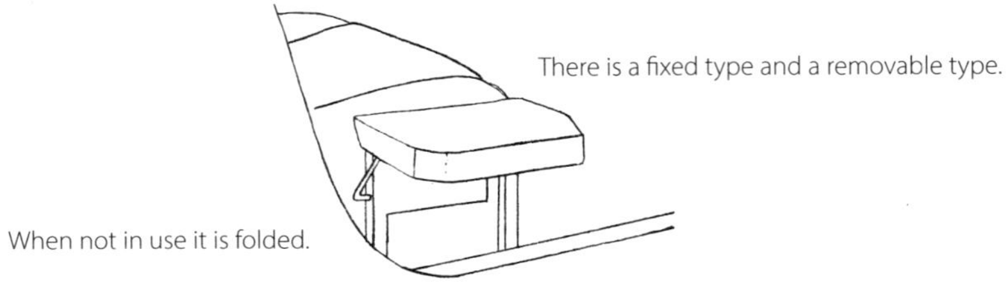

Figure 4-17 A transfer-board

When wheelchair users are driving a car with no wheelchair lift equipped, the driver needs to store his/her wheelchair in the car by him/herself. A wheelchair often weighs around 10 kg, and thus a lot of strength is necessary for the action of storing it. Shown in Figure 4-18 (a) is an electric wheelchair lift which can raise the occupant of a wheelchair to the level of the driving seat of a car and which helps the driver slide into the car. Simultaneously, the lift stores the wheelchair mechanically in the car. Models that include a wheelchair lift on board have been produced by auto-makers and these models are usually excellent in their design and functions.

Shown in Figure 4-18 (b) is a wheelchair lift with a crane attached to the roof of the car. The crane is controlled mechanically to lift or bring down the wheelchair. The arm of the crane lifts, swings around and stores the wheelchair in a storage area on the roof or the back of the car.

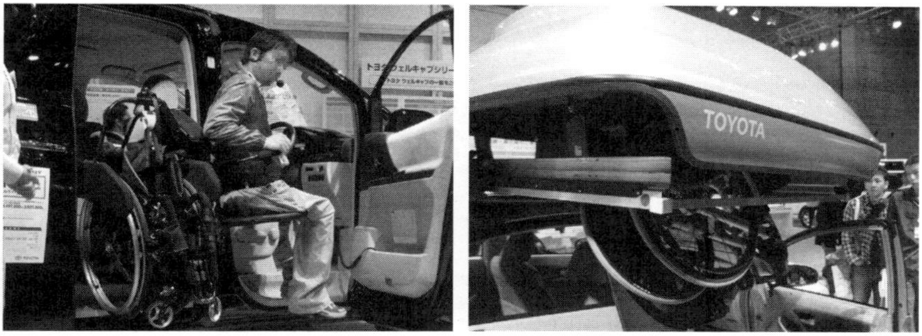

(a) Welride (Toyota Motor) (b) Welcarry (Toyota Motor)

Figure 4-18 A wheelchair lift

Various kinds of wheelchair lifts are available today, and the lift shown in Figure 4-19 is a model of a lift of the type which uses a crane arm to store a wheelchair in the boot of a car. There is another type of lift which allows the user of an electric wheelchair to transfer the chair to the driver area of a car without getting off the wheelchair at all.

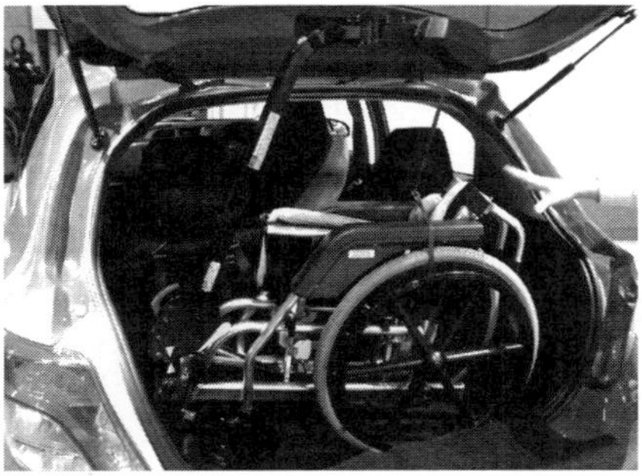

Figure 4-19 An electric wheelchair lift (Toyota Motor)

Support tools

Tools and devices have been developed in order to provide support for people who have difficulties in entering and exiting a car. A seat with a rotation and

Figure 4-20 A lift-up passenger-seat (Daihatsu Motor)

slide mechanism, for example, supports the user by rotating and sliding the seat. Shown in Figure 4-20 is a lift-up passenger-seat which moves to the outside of the car electrically. Furthermore, the height of the seat is adjustable to help the user. The driver can control the movements of the seat from inside.

There are two types of assist devices for wheelchair users, a ramp type and a lift type. These devices are often mounted on mini-buses or other transportation vehicles which are usually used when travelling between welfare institutes or hospitals and the homes of the users. The ramp type, shown in Figure 4-21 (a), is a type whereby the ramp is taken out or unfolded from the back of a vehicle. The occupant of the wheelchair moves onto the ramp while sitting in the

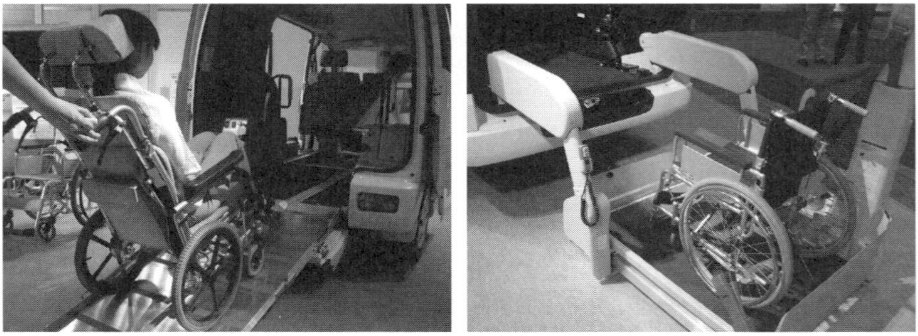

(a) Daihatsu Motor

(b) Toyota Motor

Figure 4-21 Transportation vehicles equipped with assist devices wheelchair. The occupant is then fastened by a seat belt during actual transportation for safety. The lift type shown in Figure 4-21 (b) is usually mounted on relatively large vehicles. The lift is controlled by a remote control. A square platform lifts the wheelchair up to a level where the occupant can move the chair into the vehicle smoothly.

Problems in driving vehicles experienced by persons with physical disabilities

All humans should be able to live in ways based on their own decisions in a society while their basic human rights are respected. This is a concept called the normalisation of society. As a basic principle in crystallising this concept, the idea of universal design has become widespread. This idea is applied in a variety of fields, from the design stage onwards, with the purpose of pursuing a safe and comfortable lifestyle for everyone. Furthermore, with the Barrier-free Transportation Act, the creation of a barrier-free environment has been promoted in many fields, including the public transportation system. The purpose of this is to eliminate barriers and to improve the convenience and safety of the public transportation system used by the elderly and by physically disabled persons [25-27].

Many efforts have been made in improving the public transportation system for persons with physical disabilities. However, no policy has been created regarding the driving of vehicles by persons with physical disabilities in Japan. Driving a car affords an opportunity of being independent to persons with physical disabilities, although it is true that there are a number of problems yet to be solved [28, 29].

Problems related to vehicles and driving posture

When negotiating curves in a vehicle, large movements are required in order to control the steering wheel of the car. Moreover, the centrifugal force exerted has a negative influence on driver performance. In particular, when a car is being driven on a curve and turns in a leftwards direction on the road, the body trunk of the driver actually moves in a rightwards direction. This creates further difficulties for a disabled driver when he/she tries to control the car (see Figure 4-22).

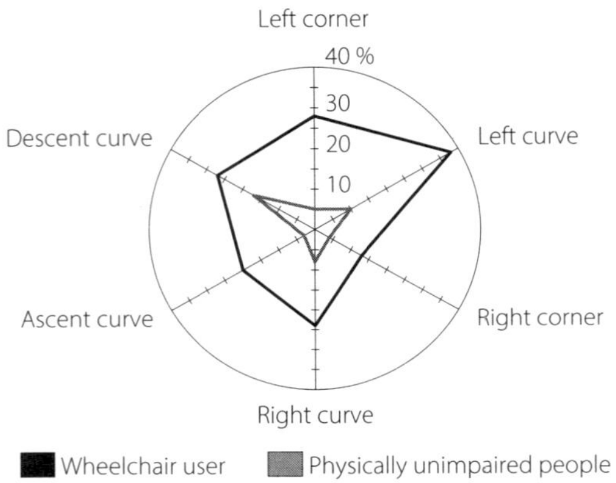

Figure 4-22 Problems caused by centrifugal force

When driving on the curve of a road, unimpaired people keep their balance by using their entire body. In contrast to this, a person with impairments to lower limbs, for example, fails to keep balance with their hips or the lower parts of their body [(30-32)]. Therefore, those persons with impairments to the lower parts of the body have to try to keep balance only by using their upper body. Thus, they have more difficulty in keeping balance if they drive on a curve fast because the force that swings their body to right and left directions increases as the speed increases. Moreover, when they are driving on a long curve, they have to try longer and harder to keep balance with their body. On S-shaped curves, the force which swings the driver's body to right and left directions switches from one side to the other at certain points, and this adds further difficulties for disabled drivers who are making an effort to keep their balance. In addition, the sharper the curves get, the more difficulty the driver has as he/she has to manage two actions: operating the steering wheel while simultaneously keeping balance with one's body. Driving conditions which include such curves put a heavy burden on the driver.

According to a study conducted by Ikeda et al., drivers who suffer severe disability in their lower limbs usually wish to fix some parts of their body to the driving seat and other parts of the car while they are driving. The parts of the body that drivers wish to have fixed in place are shown in Figure 4-23 [(33)].

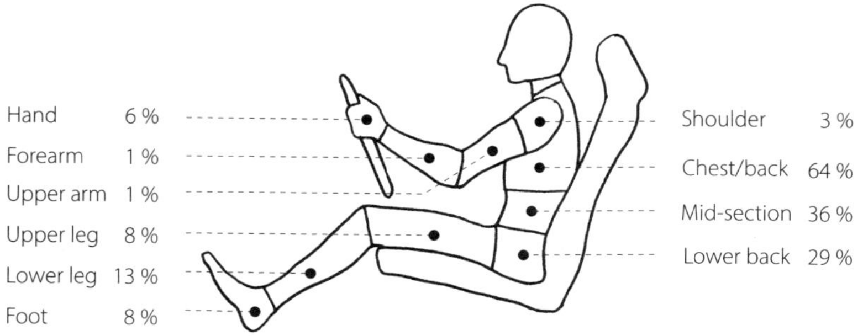

Figure 4-23 Body parts which drivers with impairments wish to have fixed in place

Figure 4-24 shows the average distance of a drivers' head movement when they are driving on curves (34). When the speed of the car reaches 30 km/h or more on a curve turning to the left, the average distance of movement of the driver's head to right and/or left increases by 2-3 times (and the average distance of movement of the driver's head in directions backwards and forwards increases by 4-5 times) over the distance of head movements by drivers in cars driven at 10 km/h on the same curve. These results indicate that when a car is driven at a speed of 30 km/h or more on a curve turning to the left, the movements of the driver's head in right-left and back-front directions increases significantly.

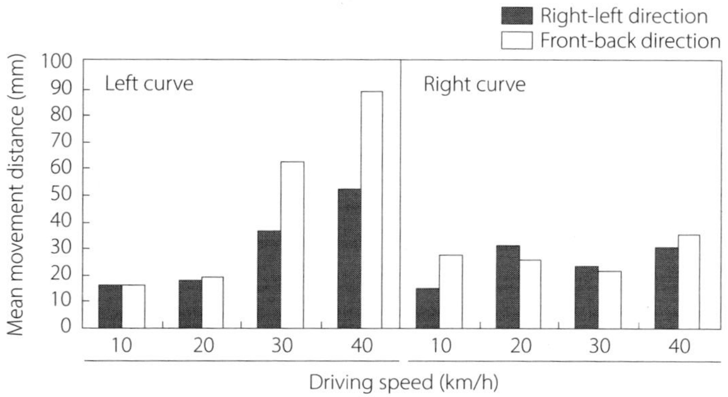

Figure 4-24 The movements of a driver's head at different speeds

Physical problems while driving

There are cases where the lower limbs of persons with impairments to their lower limbs flex unconsciously during driving and the legs have contact with the steering wheel-control and/or foot pedals. This spasticity is a symptom which usually occurs to patients with a damaged spinal cord and the flexing occurs with no relation to their conscious will. If they step on the accelerator or brake pedals unconsciously, these actions could cause a sudden acceleration or a sudden braking of the car. Moreover, medical problems could occur to a driver with physical difficulties during driving. Anaemia, for example, is one of the conditions that could develop during driving. Poor blood circulation in the lower limbs often occurs to those persons with impairments to their lower limbs during driving. This situation causes anaemia because the driver sits in the seat for a certain period of time and the blood in the body is collected at the terminal parts of the body.

Another example is a rise in blood pressure caused by the desire to go to the toilet as the desire then increases palpitation. One of the symptoms for a patient with a damaged spinal cord, when urine is collected in his/her bladder or he/she has a desire to urinate, is that blood pressure rises suddenly. This can cause headaches, diaphoresis and a worsening of spasticity. When such a condition develops while driving, of course it can cause difficulty in driving. There are many reports from persons with physical disabilities of having experienced a blood-pressure rise when they felt that they needed to go to toilet. Consequently, many disabled drivers limit the amount of food and drink they take before driving. It is difficult for wheelchair users to secure toilets when they are travelling.

In present-day Japan, rest areas called service areas are located on motorways, for example, and the toilets in the areas are built as a barrier-free zone, with parking spots for disabled persons which are built at service areas. However, often the parking spots are not clearly indicated and it is difficult to find them. In addition, often the route between the parking spots and the toilet is not built as a barrier-free zone. It is necessary to take steps to improve roads, vehicles and other traffic conditions so that persons with physical disabilities can deal with any sudden change in their physical status.

Impact of centrifugal force when driving on curves

Figure 4-25 shows the results of a study conducted by Hiroshi Ikeda et al. [35] on the impact of centrifugal force when driving on curves. As shown in the figure, when travelling speed increases, the acceleration velocity in right-left direc-

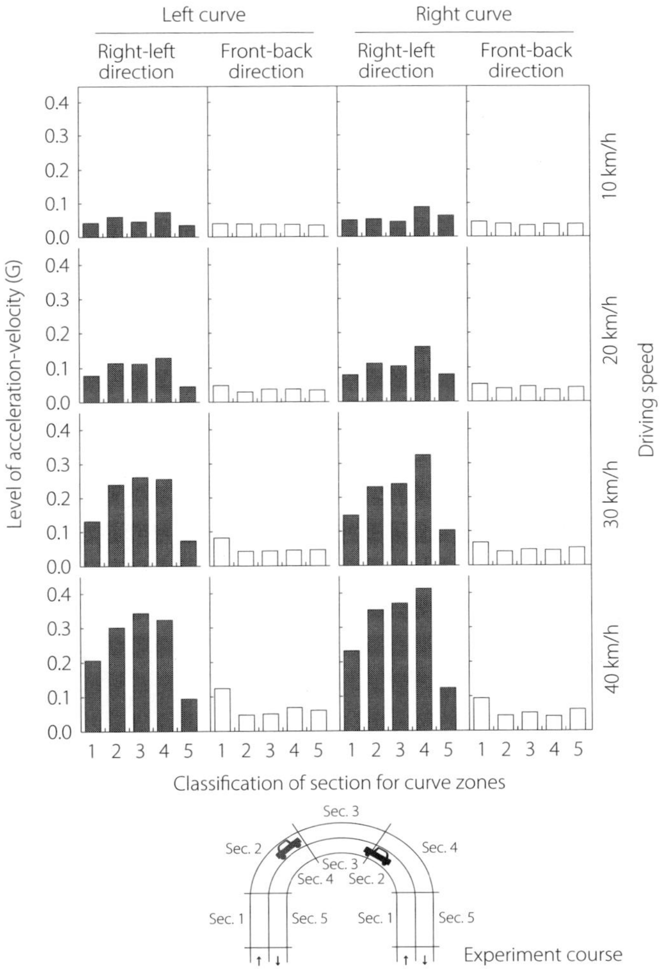

Figure 4-25 The impact of centrifugal force when driving on curves

tions changes significantly at both the start and end points of a curve zone. The impact of centrifugal force becomes greatest at the end of the curve zone when travelling in a rightwards direction. When a car is travelling on a curve, the steering wheel has to be turned back just before the end of the curve zone and, consequently, the car makes a swinging movement. The car swings more if it is driven by a person with severe disabilities in the lower limbs as he/she needs to operate the wheel by using his/her upper limbs.

Figure 4-26 shows the results of the study conducted by Ikeda et al. aimed at finding out how much muscle tension is felt by drivers when driving on curves. The level of tension was identified by measuring the amount of muscle activity in the trapezius and the extensor carpi radialis muscles of drivers. Tension in the trapezius muscle of the shoulders is related to the elevation of the elbows and the fixation of the upper body, while the extensor carpi radialis muscle in the lower arms is related to the movements which keep hand-joints and elbow-joints locked when something like the steering wheel of a car is

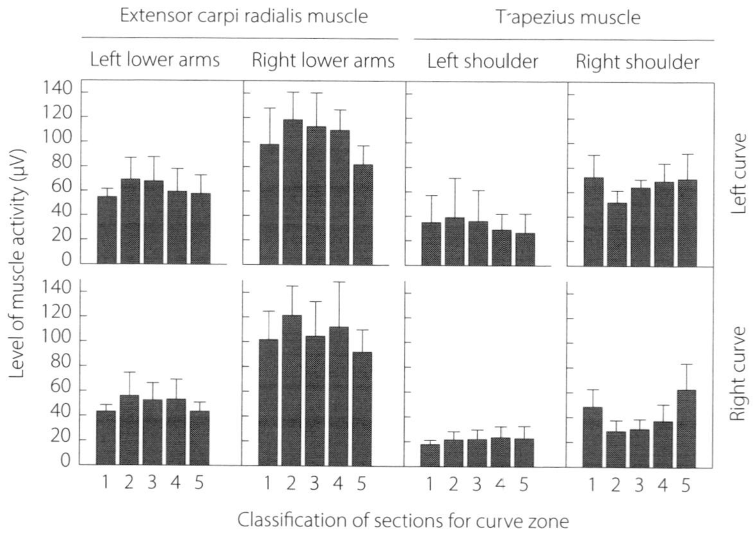

Figure 4-26 The amount of muscle activity in different parts of the body when driving on curves

being held. As the figure shows, the amount of muscle activity of the left and right forearms increases at both the start point (2) and the end point (4) of the curve zone. In comparison, however, the increase in muscle activity of the right shoulder is more significant than that of the left shoulder at the start point (2) and the end point (4) of the curve zone. When persons with severe disabilities drive a car with their upper limbs, they often use one hand for controlling the steering wheel as the other hand is being used for operating equipment such as the assist device. Under such circumstances, the body muscle gets tense as the driver tries to maintain the balance of his/her body trunk while simultaneously receiving a muscle burden from operating the car. The tension and burden on muscles cause a delay in returning the steering wheel back to normal just before the car gets on to the straight zone of the road when exiting from the curve zone. As a result, the action of turning the steering wheel becomes sudden.

Driving posture and methods for retention of the body trunk

Automobile drivers with severe disabilities in their lower limbs often find some conditions difficult to handle which are in fact normal for unimpaired people. Their disabilities cannot be compensated for only by remodelling a basic car which was originally designed for use by unimpaired people, even if such cars are remodelled by installing driving-assist devices and have their operating procedures modified. It is necessary to assess and understand the driving conditions of disabled people scientifically.

Most commonly, drivers with physical disabilities control the steering wheel with one hand while controlling the accelerator and brake with the other hand during driving. Under such conditions, controlling the steering wheel with both hands is not possible and, as a result, the centrifugal force and other forces from outside prevent the driver from keeping his/her body balance stable during driving. This condition gets especially worse when disabled drivers negotiate curves at certain constant speeds, as the driver's skill at controlling the steering wheel is affected negatively by the effort to maintain his/her balance of the body. Furthermore, the technique for keeping the body in balance is different for right and left curves, as is shown in Figure 4-27.

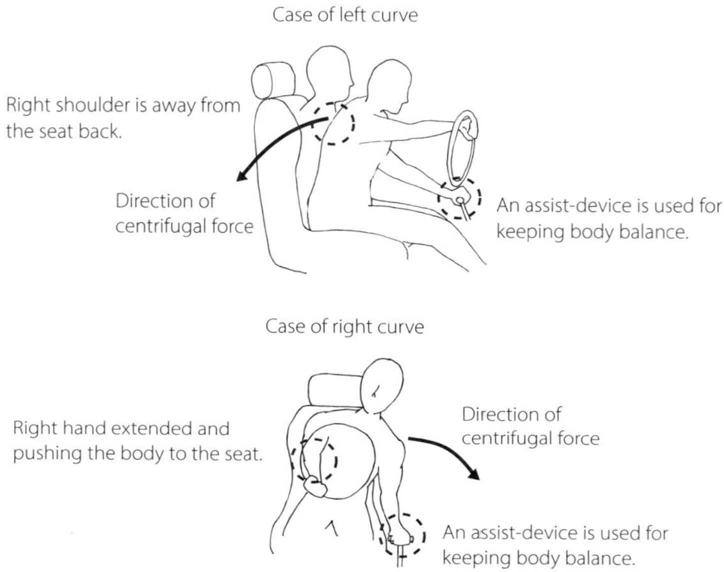

Figure 4-27 Driving posture when driving on curves

Turning the steering wheel with large movements or hitting the brakes can break down the balance of the driver's body trunk, and under such circumstances the driver has to drive the car in an unstable posture. When driving on a leftward curve, the sitting posture changes more rapidly. This situation interferes with the functional ability of the driving seat, which is mechanically designed to keep the body of the driver in a safe and stable position. Thus, the tension in the body felt by the driver becomes even stronger than the tension felt by the driver when the car is negotiating a rightwards curve. In contrast, on a rightwards curve the driver's body is in a posture where the body trunk can be controlled against the effects of centrifugal force. However, when the car is at the end of the curve zone and moves on towards the straight zone of the road, the driver loses balance because of the effort of maintaining his/her body posture against the effects of centrifugal force, and a longer time is needed for turning the steering wheel back to the original straight-ahead driving position than when the car is on a leftwards curve. Consequently, the car swings greatly at the end of the curve zone and at the point of returning to a straight road, and this increases the burden on the muscles in the neck and shoulder of the driver.

A decrease of the power to retain the body trunk in balance causes the driver to control the car in an unstable posture. Therefore, improvements in road conditions should be made: for example, by providing more space at the end of curve zones or by introducing induction strategies so that the starting points of curve zones are indicated in advance. Such improvements are difficult to make as they have to be initiated by governments and require time and money, at least under present circumstances. However, realistic and effective methods are also available. One of these methods is, for instance, the adding of safety devices to the driver's seat of existing cars. Such devices can be developed by examining closely the problem of drivers' losing body balance during driving. However, these devices have to be developed with an effort also made to minimise the financial burden on the users.

Remodelling and financial aid

Beside the hand-operated assist-devices mentioned, assist-devices for, for example, the steering wheel and direction indicators are the most common devices which are often later added to a car. However, almost all disabled drivers are using a driving seat without remodelling or adding further assist-devices, even though the existing seats usually do not satisfy the demands of the users. One of the reasons is the amount of subsidy for remodelling cars. In Japan, about 70 % of vehicle owners who have physical disabilities have paid more than half of the cost for remodelling by themselves. Shown in Figure 4-28 is a summary of the amount of money paid by individuals, made based on a survey conducted on the total amount needed to install assist devices for driving and the amount of subsidy the owners received.

In Japan, the amount of subsidy given to installing assist-devices is supplied with a certain amount of money as an upper limit. Therefore, the severer the disability of the driver, the more assist-devices are needed, and the more money has to be paid out by the individual for remodelling. Under such circumstances, people tend to remodel cars as little as possible, and consequently the remod-

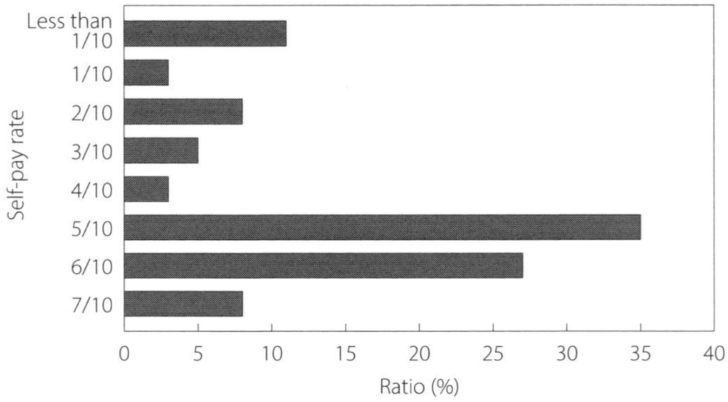

Figure 4-28 The amount of money paid by individuals to remodel cars

elled cars are probably neither comfortable nor safe.

In order to support safe driving by persons with severe disabilities, an appropriate environment should be developed so that the drivers and their cars are not affected negatively by driving speed, changes of conditions and other factors. Moreover, drivers often need to tilt their body forward or have parts of their body move away from the driving seat during driving. Taking such situations into consideration, it is necessary to design and create a steering environment in which those persons with physical disabilities can retain their body adequately in place and drive their car comfortably.

CHAPTER FIVE

Physical features of persons with a hearing disability

The time when a hearing impairment appears

Hearing impairment can be genetic (inherited) or non-genetic. Persons with congenital hearing impairment are people whose hearing impairment was present at birth. Abnormality of hearing tissues or infection with a virus (especially with measles) during pregnancy can also cause congenital hearing impairments. If persons suffer from a severe hearing impairment, and if they are unable to hear even with the help of a hearing aid, then they are unable to learn language through the medium of their ears. Therefore, they need to learn language by using other methods. Learning language without depending on hearing is difficult, and hearing-impaired people often face problems such as phonation and have difficulty in understanding complicated expressions of language.

Persons with a non-genetic hearing impairment are those people whose impairments have developed before, during or after birth. Non-genetic causes include: illness, drug side-effects, a long period of exposure to heavy sound-noise, injury to the head, sudden deafness caused by mental stress and damage to the hearing tissues. If a person's hearing is impaired after having a long period of normal hearing, the person usually has a good understanding of language and can also speak clearly. If their hearing impairment developed in infancy, however, or if it has been a long time since the development of the impairment, persons usually have difficulty in clear phonation as well as difficulty in speaking in long sentences.

Moreover, there is a case of senile hearing-disability. This is the condition where hearing ability deteriorates as people become old.

Sometimes brain damage and cerebral vascular disease are associated with a hearing disability. Persons with these medical conditions have difficulty in distinguishing among sounds, even though they can hear well. Another common symptom of hearing disability caused by ageing is a complication of dementia. Communication with such patients helps in reducing mental anxiety.

Types of hearing disability

Persons who have difficulties in hearing are called persons with a hearing disability. However, the causes and types of impairments vary, as well as the degree of the persons' hearing ability, and thus it is often difficult to distinguish or define such persons. People with a hearing disability are generally categorised into the groups of late-deafened people, hard-of-hearing people and deaf persons.

Late-deafened people usually refers to those persons who lost hearing ability after they had acquired speech language. Even late-deafened people who have no hearing ability at all can speak in many cases. Hard-of hearing people refers to those who still have some hearing ability left. Some can have normal conversations with other people by using a hearing aid, while some can hear only noise. Deaf persons refers to those people who lost their hearing ability before they acquired speech language, and their first language is usually sign language.

In modern schools for hearing impaired children, speech training is usually conducted by using a high-performance hearing aid. This training helps children learn the skills of speech, and they often grow up with no impairments in speech.

Differences in the nature of hearing loss

Figure 5-1 shows the anatomy of the ear. The ear is made of parts called the auricle, ear canal, eardrum, middle ear and inner ear, and auditory nerves inside are connected to the brain. Differences in the nature of hearing loss are deter-

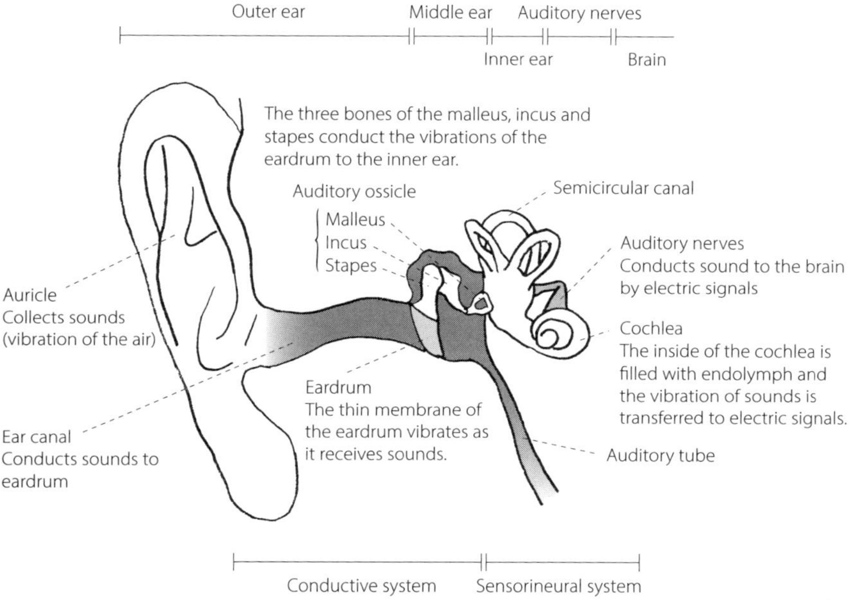

Figure 5-1 The anatomy of the ear

mined by the place where impairment occurs, and the nature of hearing loss is categorised into conductive, sensorineural and mixed hearing loss. Further details of the nature of the hearing loss are described below.

(1) Conductive hearing loss

Patients with conductive hearing loss have impairments in the eardrums, ossicles or organs which conduct sounds from the outer ear to the middle ear. The patients usually have no abnormality in the nerve system and sometimes their condition can be improved by medical treatment. Hearing loss which is caused by otitis media belongs to this category. Patients with conductive hearing loss often regain relatively good hearing ability if they use a hearing aid.

(2) Sensorineural hearing loss

Patients with sensorineural hearing loss have impairments in the organs which sense sound. This occurs when the patients have impairments between the inner ear and middle ear. Many of the patients suffer from an abnormality in the nerve system and the condition often can not be improved with medical treatment. Some patients with sensorineural hearing loss are capable of having a conversation in a quiet environment by using a hearing aid if the other speakers speak slowly. The frequency of sounds affects the degree of their hearing ability. In particular, those patients who have developed problems in hearing high-frequency sounds in infancy have difficulty in learning phonation. If persons have a severe condition, they need to learn visual communication skills, including sign language, lip reading and writing. Hearing loss which is caused by ageing also belongs to this category.

The brain analyses sound through the inner ear and auditory nerves, and then the processes of deciding whether the sound needs to be heard or not are conducted. Persons who have a hearing disability caused by functional impairments cannot regain their hearing ability by amplifying sounds with the help of a hearing aid. The quality and volume of the sounds of the hearing aid should be adjusted in detail.

(3) Mixed hearing loss

Persons with mixed hearing loss suffer from the symptoms belonging to both conductive and sensorineural hearing loss. The functions of the sound-conductive area in the middle ear can be compensated for by using a hearing aid, but the functions of the sound-sensorineural area are not always able to be compensated for by a hearing aid.

Senile hearing loss

Hearing ability deteriorates as people become older. The degree of hearing loss differs among individuals, and the deterioration begins at the age of around 40. The nature of this impairment is hearing difficulty with high-frequency sounds; this occurs as the initial symptom. The condition of senile hearing loss cannot be improved by medical operations or drug therapy. When senior adults realise that hearing difficulty is causing problems in their communication with others, they usually chose to use a hearing aid.

However, as they become older, senior adults with such hearing difficulty often loose the opportunity to have communication with other people, and often they do not realise that they have problems in hearing. Even if they are diagnosed with hearing loss, they sometimes avoid using a hearing aid because they believe that wearing a hearing aid gives an impression to other people that they are old. However, if patients spend a long period of time under such a condition of hearing loss, stimulation of the brain decreases and, as a result, their ability to distinguish among sounds and language further deteriorates.

Classification of impairments in hearing ability

(1) Classification made by the degree of impairment

The degree of hearing disability is expressed by decibels (dB), the unit used for sound pressure, and the decibel value indicates how bad a person's hearing ability is in comparison to healthy people. Table 5-1 shows the classification according to decibels of persons with a hearing disability.

Table 5-1 Classification by decibels of persons with hearing disability

Measured value	The sound description made by the patient	Degree of hearing
0-25 dB	No problem in hearing	Normal
25-40 dB	Having difficulty in hearing small voices	Slight
40-70 dB	Having difficulty in hearing normal conversation	Medium
70-90 dB	Loud sound made next to the ear can be heard	High
90 dB-	Hardly any sound is heard	Anacusis

(2) Classification made by the Physically Disabled Persons Welfare Act

Grades are given to those persons with a hearing disability in accordance with their degree of hearing ability, and the grades are stipulated by the Physically Disabled Persons Welfare Act in Japan. Persons whose level of hearing ability is 70 dB or more are supplied with an identification booklet for the physically disabled. About 360,000 people with a hearing disability are said to have been supplied with an identification booklet for the physically disabled in Japan. However, the number would be as many as 6 million people if the number were counted based on the recommendations made by the World Health Organisation (WHO). WHO states that those people whose hearing level is 41 dB or more are recommended to use a hearing aid. In Japan, persons with a hearing disability are eligible for a second-degree certification of physical disability, and persons with both a hearing disability and speech impairment are eligible for a first-degree certification of physical disability.

Table 5-2 Grade of disability for persons with a hearing disability

Grade of disability	Decision criteria
1st class	Not relevant.
2nd class	The hearing level of both ears is 100 dB or more.
3rd class	The hearing level of both ears is 90 dB and more. (Loud voice can be heard only if the sound is made directly to the auricle of the ears.)
4th class	1. The hearing level of both ears is 80 dB and more. (Loud voice can be heard only if the sound is made directly to the auricle of the ears.) 2. The best articulation score in speech sound is 50 % or less for both ears.
5th class	Not relevant.
6th class	1. The hearing level of both ears is 70 dB or more. (Conversation which is being made at a place 40 cm or more away from the person cannot be understood.) 2. The hearing level of one ear is 90 dB or more, while the hearing level of the other ear is 50 dB or more.
7th class	Not relevant.

Notes:
1. If the patient has two overlapping impairments which belong to the same class in the table, the person will be upgraded by one class. However, if the two overlapping impairments are specified in the table, the appropriate class will be given to the person.
2. If the patient has two overlapping impairments which belong to different classes, the class may be upgraded in the view of the severity of impairment.

CHAPTER SIX

Communication and transportation of persons with a hearing disability

Hearing aids and artificial ears

A hearing aid is a tool which can compensate for the hearing ability of those persons who have lost their hearing or have an impairment. When a loud sound is needed to be produced, sounds are first collected by a microphone, then amplified and finally reproduced by a speaker. This mechanism is installed in a small device, and it has been developed as a hearing device. Some of the newly developed hearing devices have a function which operates as an adjuster of sound. The function improves the efficiency of the hearing aid as the range of sounds heard by persons with a hearing disability is often significantly different among different individuals. Moreover, some aids are digitalised and they are programmed so that the device is controlled automatically in accordance with the situations they are in. This permits them to have hearing functions close to that of an unimpaired person.

An artificial ear is an instrument which compensates for the user's hearing disability by stimulating the hearing nerves electrically through a thin electrode which is implanted in the cochlea of the inner ear of the user. Sound picked up by a microphone hung on the ear of the user is converted into electric signals by a sound spectrograph, and the signals are then transmitted to an electrode in the inner ear. In contrast, a hearing aid is a device which is placed outside of the ear and it conveys a sound from the middle ear to inner ear while amplifying the sound. The sound produced by a hearing aid is closer to the sound heard by people with normal hearing ability than the sound produced by, for example, an artificial ear. On the contrary, the sound produced by an artificial ear is heard by the nerve centre of the ear, as the sound is produced while an electrode implanted in the inside of the cochlea is stimulating the hearing nerves of the ear.

When a hearing aid is put on each ear, the range of sound heard by the user expands significantly in comparison to the case where only one hearing aid is put on either the right or left ear. By using both ears, people can recognise such things as language and the movements of people's faces, and also they can analyse the distance to the sound and the direction of the sound. If people hear sound from only one ear, the distance and direction of the speakers will be difficult to recognise. Acoustic sense helps people distinguish what they want to hear among many different sounds. This function, so-called "cocktail-party effect", becomes active when both ears are used. Hearing with both ears can also reduce the amount of fatigue felt by a person because even small sounds can be picked up easily.

Types of hearing aids

Basically, four types of hearing aids are available as described below. The most popular ones are the type for inserting into the ear and the type for hanging on each ear. There are other types of hearing aids beside the four types described below. Some aids are embedded with a FM receiver which is effective in noisy circumstances, while some aids can compress frequency so that the treble consonant is easily heard.

(1) Hearing aid for inserting in the ear

This aid can be put into the inside of each ear and it is not easy to be seen from outside. This type is usually tailored in accordance with the shape of each person's ear and the different hearing-ability of users. Once it is made, it fits well to the ear shape of each user and therefore it will not fall out from the ear easily. Thus, the users find it very comfortable. The acoustic quality of

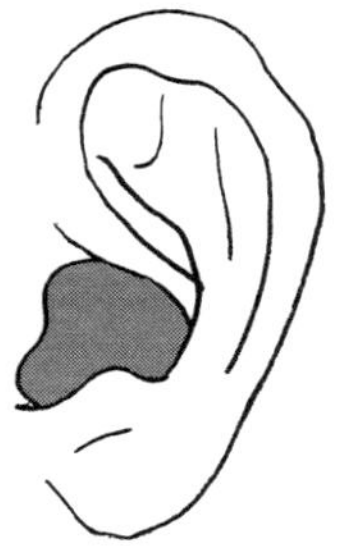

Figure 6-1 A hearing aid for inserting in the ear

this type of hearing aid is very close to the sound heard by unimpaired people, since the sound produced by the hearing aid is conveyed directly into the ear-hole. However, the price is higher than other hearing aids, and this aid usually has fewer functions than other hearing aids. Another disadvantage of this type is that when the aid is put very tightly into the ear, the user can hear the echo of his/her own voice, and also when the aid is inserted relatively loosely, the user hears a sound as if the aid is howling.

(2) Hearing aid for hanging on the ear

This is a type that hangs on each ear, so that the aid is set behind the ear. A variety of size, models and price of hearing aids is available, and they match different levels of hearing disability. The volume is adjustable either manually or automatically. This type of aid for hanging on the ear comes with a variety of options. However, they also have a few disadvantages. One is that sweat from the user can get into the device and cause mechanical problems. Another disadvantage is usually felt by people who wear eyeglasses in everyday life. This type of hearing aid is designed similar to glasses, and the users often find that it is not comfortable to wear the hearing aid and glasses at the same time. However, in addition to the merits already described, there is one more significant merit to this type. This type of hearing aid produces less howling compared to, for example, the hearing aid which is designed to be inserted in the ear. This is because the microphone which picks up sound is designed to be kept away from the ear-hole. The type of hearing aid which hangs on each ear has become the standard type world-

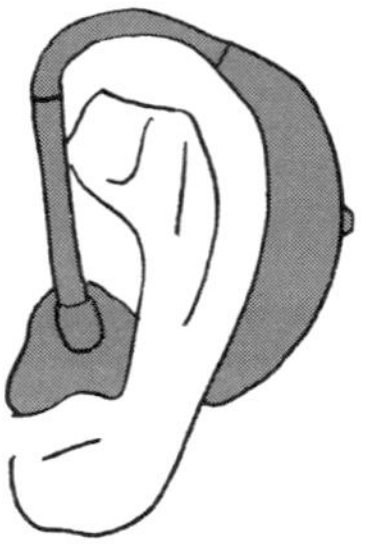

Figure 6-2 A hearing aid for hanging on the ear

wide, though the type inserted in the ear is still popular in Japan.

(3) Hearing aid for putting in a pocket

This is a hearing aid of which the shape is similar to a portable radio with an earphone. The sound produced from the speaker of the aid is carried directly to the inside of the ear and this helps producing a good quality of sound. The cost of the aid is not expensive. It also has the advantage that operation of the aid is easy because the body of the aid is big. However, some people find that the cord between the body and the earphone is annoying and that the rubbing sound created while the user is walking is noisy. The design and specification of this type are seldom renewed, and an analogue circuit

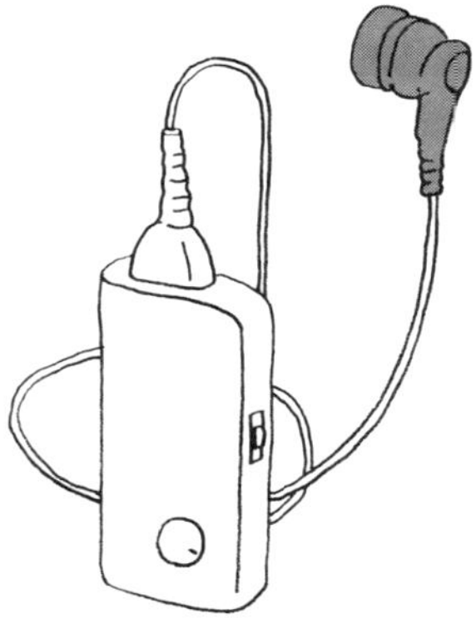

Figure 6-3 A hearing aid designed to be put in a pocket

is usually embedded in the pocket-type of hearing aid. Both hearing aids of the type inserted and of the type hanging on the ear are more suited for users who play sports.

(4) A spectacles-type of hearing aid

Two types are available. One type is a hearing aid which is designed to be set onto each temple tip of a pair of spectacles after their purchase, while the other type is designed to be embedded into the temple tip of a pair of spectacles before purchase. The glasses and hearing aids are put together; thus the aids do not easily fall off and they are not easily seen from outside. However, spectacles and hearing aids cannot be separated—or cannot easily be separated—so that when the lenses of the spectacles require a change, the aid also has to be changed or readjusted.

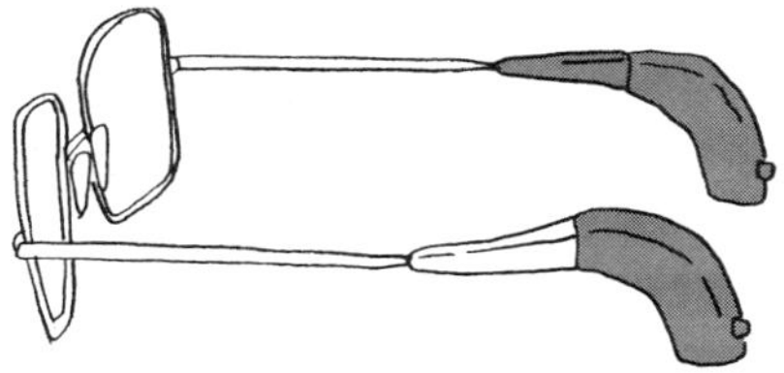

Figure 6-4 A spectacles-type of hearing aid

Digital hearing aid

With an analogue hearing aid, the sound signal collected by the microphone is amplified and produced from an earphone. In contrast, with a digital hearing aid, the sound signal is at first converted to digital sound, and amplification and other complicated processes are conducted. Then the digital sound is converted to a sound signal again, and finally, the sound is produced from an earphone. Processing of a digital signal can eliminate noise and distortion of sound, and thus the sound produced from a digital hearing aid is very clear. A digital hearing aid has a mechanism to pick up only conversations among different sounds and noises, while emphasising the treble consonant to help hearing.

The function for eliminating noises has been improved and newly devel-

Classification of hearing loss	Normal (0-25 dB)	Slight degree of hearing loss (25-50 dB)	Intermediate degree of hearing loss (50-70 dB)		High degree of hearing loss (70-90 dB)	Severe degree of hearing loss (90 dB or more)
	The person can hear well	The person has difficulty in hearing a small voice.	The person finds difficulty in having normal conversation.	The person has difficulty in hearing unless the sound is produced next to the ear.	The person has difficulty in hearing even loud voices unless the speaker speaks next to the ear.	The person can hardly hear anything.
Type for inserting in the ear	←	—	—	—	→	
Type for hanging on the ear		←	—	—	—	→
Type for putting in a pocket		←	—	—	—	→
Type for spectacles		←	—	—	→	

Figure 6-5 The degree of hearing loss and suitable hearing aids

oped digital hearing aids can eliminate ambient noise significantly. However, it has been found that the elimination of ambient noise makes the user uncomfortable or even scared, and that ambient noise is necessary as a source of information for the users.

Wearing hearing aids on both ears

Wearing a hearing aid on or in each ear can activate the hearing-related functions more than when an aid is worn by only one ear. Hearing ability improves significantly, and the burden put on the ear of the users is reduced, when aids are worn on both ears. In America, about 68 % of all hearing-aid users wear two hearing aids in or on each ear. The number is increasing, though gradually, in Japan, too.

The effects of wearing aids on or in both ears are:

(1) It is easy to distinguish between what is necessary and what is not among different noises. It is easy to hear conversations in noisy circumstances.
(2) It is easier to distinguish and recognise the direction and position of the sound being produced, as well as the depth of sound.
(3) Sounds heard are closer to what is heard by an unimpaired person.
(4) Less tiredness is felt by users, since the users can hear even small sounds.

Yasumasa Arai [36] has pointed out that both the right- and left-sides of brains are activated more when a hearing aid is worn in or on each ear at the same time, and this helps the users to have more complicated and sophisticated communication. Much of the sound collected by an ear is conveyed to the side of the brain opposite to the side of the ear. For example, information which enters the right ear is conveyed to the left-side brain. There is a cadence in conversation, and the cadence helps conveying, for example, emotions which cannot be expressed through language. The meaning of language is processed by the left-side brain, while emotions and other information are processed by the right-side brain, as shown in Figure 6-6.

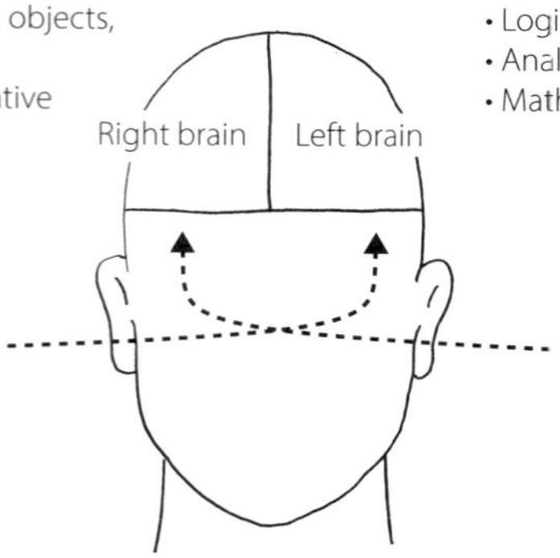

Figure 6-6 The actions of the right and left brain

Purchase of a hearing aid

Consultation with one or more otolaryngologists is necessary in the case of purchasing a hearing aid. It is necessary to have check-ups to find out the status of one's hearing. It is also important to find out what the cause of the hearing loss is and how much hearing ability is left. Based on the diagnosis, otolaryngologists can advise the patient as to what the most appropriate hearing aid will be, and they may even offer advice on which shops are better for the purchase of the hearing aid. Usually, certified hearing-aid technicians will provide appropriate advice in the shops or with the dealers of hearing aids. Those people in Japan who need to purchase a hearing aid are, in principle, recommended to buy a hearing aid in affiliated stores of dealers' associations, or in special shops which satisfy certain conditions about facilities as well as having certified staff. Purchasers need to think about battery changes or the sort of fine adjustment which is unavoidable in the future.

Figure 6-7 is a mark which indicates those shops certified by the Association of Technical Aids (ATA) of Japan, proving that they have passed an examination of qualification. Shops which have been certified by ATA are required to have a screening every five years.

Figure 6-7 The mark indicating that a shop is certified by ATA

History of the hearing aid

The hearing aid has a long history and dates back to the time when people were using animal horns and shells to amplify the sounds that enter ears. Such primitive hearing aids were improved and, eventually, a hearing aid called an ear-trumpet (shown in Figure 6-8) was introduced. The ear-trumpet is used by being inserted in the ear. Ear-trumpets were produced in a variety of shapes and types. Beethoven, for example, was using this as a hearing aid in his later life. The mechanism and design of the ear-trumpet were improved over the years and it was industrialised by Frederik Charles Rein in London in around 1800 (37).

Ear-trumpets were commonly used in the 18th and 19th centuries. By then, ear-trumpets were made of brass or thin metal materials. Hard rubber which was painted black was used to cover the aid, in order for it to blend in with the colour of clothes people were usually wearing around the time. A large-size ear-trumpet could produce a sound of 40-50 dB and this acoustic gain was fa-

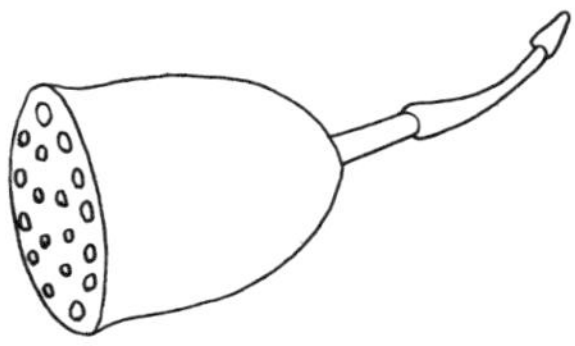

Figure 6-8 An ear-trumpet

voured by many people. The design was improved, and a famous painter, Ludwig Kirchner, designed an ear-trumpet in 1684 which is housed in a museum today [38].

In Japan, a well-known painter, Koukan Shiba, designed a trumpet-style hearing-aid called "ear-glasses" under the supervision of a Dutch person in 1813 [39]. Figure 6-9 is an advertisement for the ear-glasses. The hearing aid was termed "ear-glasses" as a contrast with the term "eye-glasses."

Figure 6-9 An advertisement for the hearing aid designed by Koukan Shiba
Source: Kobe-city Museum in Japan

The production of the modern type of hearing aid began in the 19th century. Modern hearing aids were produced by using telephone technology, and allegedly, the history of electric hearing-aids started in 1876 when Alexander Graham Bell invented the telephone. Until then, hearing aids were not effective for persons with a severe hearing disability, but through the invention of hearing aids which used batteries and a carbon microphone these devices made a difference to the lives of persons with severe hearing disability. An electrical transistor was invented by William Bradford Shockley in 1947 and this was a breakthrough in the technology of the hearing aid. Since then, hearing aids have been made smaller.

In Japan, a hearing aid installed with a carbon microphone was produced in the 1930s. The production of the evacuated-tube hearing-aid began in 1948, and hearing aids started to be sold on the market. In around 1955 transistor-type hearing-aids were made and the use of hearing aids increased rapidly.

Methods of communication

When persons with a hearing disability have communication with other people, they need special communication methods. The methods differ depending on the age when they lost their hearing ability, how much hearing ability remains, language ability, lip-reading ability, locutionary ability, their educational history and other factors. It is generally thought that all people with a hearing disability are capable of using sign language, but actually many of them cannot use it. Persons with a hearing disability need to have more than one communication method if they want to obtain satisfactory information. For example, if they are having a conversation, they need to do lip-reading while listening, with hearing aids, to what the other person is saying. If it is in a lecture setting, for example, they need to pay attention to a sign-language interpreter while reading what is written on a screen. In this way, they try to obtain as much information as possible.

Some communication methods which are conducted by persons with a hearing disability are described below.

(1) Sign language and manual alphabet

In Japan, the sign language is composed of two styles; Japanese Sign Language and Manually Coded Japanese. Japanese Sign Language was developed originally as a communication method among persons with a hearing disability. Communication is made by using hand movements and facial expressions. Japanese Sign Language has its own grammar and vocabulary, and using the sign language together with basic Japanese language is not easy. Japanese Sign Language is considered separate from basic Japanese language. In contrast, Manually Coded Japanese is a sign language of which signs are consistent with the words and grammar of basic Japanese language. When Manually Coded Japanese is used in communication, the vocabulary of basic Japanese is interpreted word by word. However, Manually Coded Japanese is not appropriate when the users are having a complicated conversation. Today, the Japanese Sign Language is gaining more popularity in Japan. A manual alphabet is used when the Japanese alphabet, the

English alphabet and numbers are needed to be expressed. Fingers are used in this method, and it is used as a supplement to sign language.

(2) Lip-reading

Lip-reading was invented by Alexander Melville Bell, one of the great teachers of physiological phonetics in the 18th century. It was introduced to Japan at the beginning of the Japanese Showa period (1926-1989). Lip-reading is a method by which the users can understand what other people say by observing the movements of the speaker's lips and tongue and their facial expression. Understanding conversations only by using lip-reading requires training and a natural gift. Those persons who lost their hearing ability after they had acquired speech language, as well as persons with hearing loss, use this method to fill in the gaps of what they cannot hear. However, some people speak very quickly and some people open their mouths only a little, so understanding what others say only by using lip-reading is very difficult. Dull sounds in words are especially difficult to pick up correctly, and it is said that the accuracy rate of picking them up is only around 30-40 %. If complicated technical terms are used, the rate decreases further. It is helpful if speakers say words slowly while looking directly at the face of the person who is lip-reading.

(3) Writing

There is a way of having communication with persons with a hearing disability by writing. It is a method where people write words on something such as a sheet of paper or one's palm, instead of producing sound or using the lips. Writing is a method with high accuracy in communication between persons with a hearing disability and unimpaired persons. It is useful in acute situations, since the necessary information can be conveyed correctly.

(4) Transcript of a summary

This is a method to interpret the contents of a speech or a conversation, by using a transcript of a summary made by interpreters who have normal hearing ability. For example, in a lecture, what the speaker says is summarised by some interpreters and they write the summary on an overhead projector so that persons with a hearing disability can read the lecture content. Personal computers are also useful for writing down such information.

(5) Other methods

Facsimile telegraphs, text messages on mobile phones, electric mails on personal computers and other forms of visual communication are also useful and common ways of communication for persons with a hearing disability. The facsimile has become established as an effective way of communication for persons with a hearing disability. Having conversations by using computer on-line services and exchanging text messages by using mobile phones and internet services have become more common among those persons with a hearing disability. A TV telephone is also in practical way for them to communicate.

Information obtained visually

Visual information is very important for persons with a hearing disability who cannot obtain much, or any, information from hearing. When TV programs and movies are subtitled, they also can enjoy them. In America, approximately 95 % of TV programs are subtitled, while only a few programs are subtitled for persons with a hearing disability in Japan. As a result of campaigns aimed at promoting TV communication accessibility in Japan, which was initiated by hearing-impaired persons, a provision for adding subtitles passed through the Japanese diet. Since then, TV stations have started taking the addition of subtitles to TV programs more seriously. In present-day Japan, a decoder for teletext broadcasting or a TV with the decoder built in is needed in order for persons with a hearing disability to enjoy TV programs with subtitles. Since the speech-conversion system was introduced, some TV programs are subtitled for persons with a hearing disability on a real-time basis today.

The methods of supplying news information to persons with a hearing disability include close-captioned TV programs, visual information radio, news services in letters on mobile phones, and home-pages on the Internet. In modern Japan, electric bulletin boards are used for supplying information related to bus, train and other transportation systems, and persons with a hearing disability find it very useful when they need to know when to get off a bus, for example. Services of rental-videos with subtitles are also available for persons with a hearing disability today.

Disadvantages in everyday life

It is difficult to know whether a person has a hearing disability or not just by looking at them, and this creates disadvantages for them in everyday life. Some disadvantages felt or found by the persons with a hearing disability are described below.

(1) It is not easy for others to notice that their hearing ability is impaired.
Persons with a hearing disability often cannot realise they are being talked to, for example, unless the speaker taps or touches some part of their body. Many of those persons with a hearing disability have experiences of being misunderstood where others think they are pretending that they cannot hear, or that they have difficulty of understanding, since they look normal on the outside. In particular, if such a person has both a hearing disability and a speech impairment, their communication is extremely limited, and unfortunately such difficulties and disadvantages are not understood by many people. In spite of facing such disadvantages every day, they are sometimes treated and abused very badly.

(2) They cannot notice or realise that some information is being broadcast, or that their name is being called.
In the case of waiting for their name to be called in places like a doctor's clinic and a bank, they often do not notice even after their name has been called. If breaking news on, for example, an earthquake is broadcast on TV, they do not notice the news unless they actually look directly at the TV screen.

(3) They cannot judge the situations around them through sounds.
People usually can judge what is happening around them, or to them, from the sounds they hear. However, persons with a hearing disability cannot make such judgements, and thus they cannot respond to a situation quickly. They often lack the information they need to have for their survival.

(4) Inadequate communication methods are used by other people.

As a way of having communication with other people, some persons with a hearing disability use, for example, sign language, while others may use writing. The communication methods vary depending on individuals. If the communication method used is not adequate for, for example, a person whom you wish to help, you may find that you cannot communicate with him/her at all. Finding a suitable communication method is thus also important.

The environment for driving cars

Persons with disabilities should be able to participate in social activities in the same way as unimpaired people. In America, for example, since the establishment of the Americans with Disability Act (ADA), the opportunities provided for persons with a disability have increased. In Japan too, since the Barrier-free Transportation Act and Heart Building Law were established, buildings and places related to public transportation systems have been created as barrier-free zones. Participation by persons with disabilities has been promoted globally, based on the normalisation principles of society.

However, not enough recognition and understanding have been given to the driving of cars by persons with physical disabilities. Currently in Japan more assist tools and devices are being developed in order to support people with severe disabilities in lower and upper limbs, and some driver tools and devices are now available, depending on the types of disabilities people have. However, the support made for persons with a hearing disability is not yet sufficient and using a hearing aid is the only method available to them so far. Even for the system related to acquiring a driver's license, the views of the government and of persons with a hearing disability are different. For instance, visual information is very important when persons with a hearing disability drive cars; therefore, they demand a systemic revision in order to improve the driving area. Meanwhile, though the government is considering reviewing the system for a driver's license, common ground has not been found yet between the government and hearing-disabled users of cars..

It is generally thought that it is dangerous for persons with a hearing disability to drive a car. However, according to Japan's National Centre for Persons with Hearing and Language Disabilities, which is the predecessor of Japan's National Rehabilitation Centre for Persons with Disabilities, there is no scientific evidence for this [40]. As one of the ways of providing technical support for persons with a hearing disability, Shinji Miyazaki et al. [41] pointed out the possibility of developing a device which can supply information to those drivers with a hearing disability. With this device, important information relating to driving can be detected among traffic noises and the noise is converted to light or vibration stimuli. Today, some companies and institutes are conducting research and development on a device which can inform the drivers of oncoming emergency-motor-vehicles by using vibrations. However, this is still not beyond the R&D phase yet.

History of the campaigns for the acquisition of a driver's license

The history of decision-making on issuing a driver's license to a person with a hearing disability started at the end of 1951, according to "Japanese DEAF news", a newspaper published by the Japanese Federation of the Deaf. It states that it was decided that "a driver's license will be issued to persons with a hearing disability" with conditions which included installing a back mirror and always having a fellow passenger in the car who had a driver's license. In 1954, however, "the deaf and mute" were added as candidates for the revocation and suspension of driver's licences in the enforcement order of road traffic regulations which are the predecessor of the Road Traffic Law in Japan.

Shigeo Takagi, chairperson of the Japanese Federation of the Deaf, expressed his opposition to this enforcement order by travelling 5,000 km in a light-size car with two people who had a hearing disability safely and successfully, in order to prove that there are no problems with the driving ability of persons with a hearing impairment. He also made a round of visits to the Tokyo Metropolitan Police Department and the Ministry of Welfare (presently the Ministry of Health, Labour and Welfare), and lobbied for the rectification of the enforcement order. This marked the beginning of the campaigns for the acquisition of a driver's license for persons with a hearing disability.

Table 6-1 History of the campaigns for the acquisition of a driver's license for persons with a hearing disability

At the end of 1951	A driver's license is to be issued with conditions which include installing a back mirror and always having a fellow passenger who has a driver's license in the car.
1954	In 1954, "the deaf and mute" were added as candidates for the revocation and suspension of driver's licences in the enforcement order of road traffic regulations which are the predecessor of the Road Traffic Law in Japan.
1958	Shigeo Takagi, chairperson of the Japanese Federation of the Deaf made a 5,000 km-trip with two people who had a hearing disability. This trip was completed successfully without violating of any traffic rules. A round of visits to Tokyo Metropolitan Police Department and the Ministry of Welfare was made by Shigeo Takagi and he lobbied for the rectification of the enforcement order.
1960	The name of the enforcement order of the road traffic regulations was changed to the Road Traffic Law. The words of "the deaf and mute" were changed to "the persons who are unable to hear and the persons who are unable to speak." The status of issuing a driver's license was changed from an absolute refusal to the delivery of the license with some conditions. The conditions included that the persons who wanted a driver's license must be able to hear 90phon of sound at a distance of 10 m from the sound.
After 1965	The idea that "a hearing aid is not good" became common in Japan, and many driver's licenses were taken away from drivers with a hearing impairment.
1967	Mitsuo Toishita, a 25-year-old Japanese person with a hearing disability, was arrested for driving without a license. In December, he refused to pay the fine by following the summary order, but instead he demanded an official trial.
1968	The Japanese Federation of the Deaf made the decision fully to support Toishita in the trial.
1972	The Japanese Federation of the Deaf called for support from its signatories, and they carried on campaigns of collecting signatures on a petition.
1973	The Japanese Federation of the Deaf formed "a committee calling for the revision of Article 88 of the Road Traffic Law which stipulates regulations for persons with hearing and language disabilities", in cooperation with related organisations, including parental groups of children with a hearing disability and the principals of schools for the deaf. On December 19th, 30,000 signatures on a petition calling for the revision were passed to the House of Representatives. The National Police Agency notified municipal police departments that the use of hearing aids was permitted during audibility tests and the term "persons who are unable to speak" was virtually deleted. For the first time, a budget was allocated to The Ministry of Health as a cost of the development of conducting experiments on the safe operation of cars for persons with hearing and language disabilities.
2001	A bill for the partial amendment in the Road Traffic Law was passed by a vote in the Lower House plenary session. The causes for disqualification stated in Article 88 of the Road Traffic Law, where a driving license is not issued to persons who are unable to hear, were repealed.
2002-	The National Police Agency entrusted a research study on the relation between hearing impairments and the safe driving of cars to private research companies, using the national budget.
Currently	Some companies are developing devices which change the alarm-horn of emergency vehicles to lights and vibrations and convey them to drivers with hearing impairments. The National Police Agency is studying Article 23 of the enforcement regulation of the Road Traffic Law and putting the final touches on deciding their direction in the future.

A special mark for persons with hearing disabilities

Shown in Figure 6-10 is a mark which should be displayed in a car driven by a person with a hearing disability. This mark indicates that the driver of the car is granted a driver's license with some conditions. The mark was introduced in 2008 following a review of the causes for disqualification of persons with disabilities. The review was conducted together with an amendment to the Road Traffic Law. The drivers are required to have the mark in their cars while driving, and driving without the mark is a violation of the law. However, if persons with a hearing disability are granted a driver's license with a condition which requires them to wear a hearing aid while driving, they do not need to display the mark in the car. In addition, all drivers in Japan have a responsibility to protect vehicles with the mark indicated, and if unimpaired drivers act badly, including attempts to cut into the small space in front of a car with the mark or driving too close to the car, they will be fined according to rules stipulated in the Road Traffic Law.

Figure 6-10 The mark for drivers with hearing disabilities

Figure 6-11 indicates that the person who has or shows the mark has a hearing disability. It is not usually clear whether a person has a hearing disability or not by their looks, and this has caused misunderstandings, disadvantages and anxiety to persons with a hearing disability. When the mark is shown, it is important to understand that the person displaying it has a hearing disability and to seek suitable communication methods.

Figure 6-11 The mark called an "ear mark" in Japan

A license system in relation to acoustic sense

In Japan, the causes for disqualification of persons with disabilities from driving have included people with psychological illness, intellectual disabilities, epilepsia, severe visual impairment and persons who are unable to hear or speak. However, sections relating to persons who are unable to hear or speak were abolished from the causes above according to the revision of the Road Traffic Law of 2001. Before 2012, drivers with a hearing disability were allowed to drive only a basic passenger car (which excludes a goods wagon), and a driver's license was granted only to persons who passed a hearing test where 90 phon of sound could be heard at a distance of 10 m from that sound. However, the regulation was amended in 2012, and they are now permitted to drive a basic goods-wagon as well if the vehicle is equipped with an auxiliary mirror and indicates the mark for persons with hearing disabilities. They are also permitted to drive a 50 cc-engine motorcycle, a basic motorcycle and a large-sized motorcycle without having an auxiliary mirror and the mark for persons with a hearing disability on the vehicle.

With regards to obtaining a driver's license, many persons with a hearing disability are calling for a revision of the obligations to wear a hearing aid and pass an aptitude test. They argue that they could drive safely without information obtained from their ears as long as they have visual information. They also point out that a hearing disability is not included among the causes of disqualification for the acquisition of a driver's license in many countries. In addition,

Table 6-2 The types of vehicles which can be driven by persons with a hearing disability

Type of vehicles		Up to the present	Currently
Basic cars	Passenger cars	○*	○*
	Goods wagon	×	○*
Motorbike		×	○
Small-sized special motor-vehicle		×	○
Large-sized motorcycle		×	○
Basic motorcycle		×	○

Note: * It is a condition that drivers must display the mark for persons with a hearing disability and install an assist mirror on the vehicle used.

some persons with a hearing disability prefer having tests on the ability to collect visual information to tests on their hearing ability. In the same way as unimpaired people, driving a car is an important way to participate in social activities for persons with a hearing disability. Unfortunately, though persons with a hearing disability have a legal obligation to wear a hearing aid during driving, many of them do not use a hearing aid for such reasons as the annoying noise caused by the aid.

Difficult traffic conditions for persons with a hearing disability

Persons with a hearing disability usually find traffic conditions such as streets where visual information is overflowing and roads with blind corners very difficult to drive. They have to rely on the information collected from their eyes during driving. If they cannot get a picture through visual information, they cannot understand what is going on around them. When unimpaired people are driving on roads with poor visibility, they can use the horn to have contact with other drivers if they need to do. However, that is not possible for persons with a hearing disability, and they often feel anxiety when they are driving on roads with poor visibility. Very noisy roads also make driving for persons with a hearing disability very difficult. Moreover, they find it difficult to drive on streets with unattended audio assistance, because, though they often find the audio sound difficult to hear, they cannot have other ways of having communication there in order to receive assistance.

Some examples of road conditions under which persons with a hearing disability find it difficult to drive are listed below.

(1) Busy roads and intersections with poor visibility
(2) Roads with many sharp curves, mountain trails and lanes
(3) Narrow streets, T-intersections and one-way streets
(4) Roads with no crossing and traffic signals
(5) Tunnels on motorways and roads with noisy construction work
(6) Streets with unattended audio assistance

Information gathering by using acoustic and visual senses during driving

Generally, people find that sound which comes from their back is more difficult to hear than the sounds from their front, their left or right, and not many people can distinguish among different sirens made by different emergency vehicles. A surprising number of people do not notice the sirens of approaching emergency vehicles until they come very near them. The effect and comfort of hearing aids vary depending on the users, and hearing aids can cause headaches and other negative factors.

Information gathering by vision is more difficult at night than in daylight hours. Darkness prevents people from finding or detecting necessary information. Moreover, less auditory information is available at night and this is likely to cause accidents and other dangerous situations.

Use of a hearing aid during driving

Persons with a hearing disability are obliged to wear a hearing aid during driving, but many of them have had one or more experiences of driving a car without wearing a hearing aid. This is partly because many people consider that visual information is good enough for driving and auditory information is not necessary. Persons with a hearing ability are often very good at grasping information about what is going around them by observing, for example, movements of other vehicles. They feel that the information is adequate for driving. In addition, the sound and noise from a hearing aid sometimes make drivers feel sick during driving. Driving with a hearing aid could cause drivers headaches, and some drivers need to adjust or take off their hearing aid while they are driving.

Dizziness is also often felt by drivers when they are driving with a hearing aid. According to a study by Haruhiro Katoh et al., drivers feel dizziness more often when they are driving at the speed of 100 km/h or more on motorways [(42)]. Generally, when people are driving a vehicle, 95 % or more of the information is gathered from vision. Moreover, the field of vision of drivers gets narrower and the focus is on the things in the distance [(43)]. This reduces the amount of visual information around the drivers. These difficult conditions are then combined with physical phenomena, such as the view seeming to approach drivers much faster during high-speed conditions than on normal roads. These factors cause—and become—unusual stimuli for drivers and make them feel dizziness.

CHAPTER SEVEN

Physical features of persons with visual impairment

Types of visual impairments

Persons with visual impairment are people who have difficulties in everyday life because they have no sight or have poor visual performance. Their visual acuity or their condition with regards to the construction of a visual field does not

Table 7-1 Different classes of persons with visual impairment

1st class	A person whose total binocular vision is 0.01 or under.
2nd class	A person whose total binocular vision is 0.02 or over, but 0.04 or under. A person whose view from each eye is within 10 degrees, and 95 % or more of the visual efficiency is lost in the binocular view.
3rd class	A person whose total binocular vision is 0.05 or over, but 0.08 or under. A person whose view from each eye is within 10 degrees, and 90 % or more of the visual efficiency is lost in the binocular view.
4th class	A person whose total binocular vision is 0.09 or over, but 0.12 or under. A person whose view from each eye is within 10 degrees.
5th class	A person whose total binocular vision is 0.13 or over, but 0.2 or under. A person for whom 50 % of the view is lost in binocular view.
6th class	A person whose visual acuity in one eye is 0.02 or under, and whose visual acuity in the other eye is 0.6 or under, while the total binocular vision is 0.2 or more.

Note: Visual efficiency is one of the methods to assess the impairment degree of visual function. The visual efficiency of unimpaired persons is 100 %, while the visual efficiency of a person who has no sight is 0 %.

improve beyond a certain degree, even if they wear glasses or if they conduct other methods for correcting or improving their visual ability. People who have poor vision but whose vision improves by, for example, wearing glasses, do not belong to the category of persons with visual impairment. The Physically Disabled Persons Welfare Act, which was implemented in 1995, defines those persons with visual impairment as follows:

Visual disability is divided broadly into the three categories of impaired eyesight, visual field defect and defective colour vision. According to the Physically Disabled Persons Welfare Act, those persons with impaired eyesight and those persons with a visual field defect—but not persons with defective colour vision—are certified as persons with physical disabilities and granted an identification booklet for the physically disabled.

(1) Impaired eyesight

Impaired eyesight is divided broadly into two categories of complete blindness and weak sight. The number of persons who are partially sighted accounts for more than 70 % of the total number of persons with impaired eyesight, and that number is significantly higher than the number of persons with complete blindness. Persons with complete blindness cannot even sense the presence of light. The visual acuity of those persons who belong to the category of weak sight does not improve even if they attempt to correct their vision by wearing glasses or contact lenses. Their total binocular vision does not reach 0.3 or more, even with glasses or contact lenses. The causes and symptoms of the persons with weak sight vary, and their ability to see also varies among individuals. The weather and their psychosomatic states can affect their ability to see, and the power and status of vision change even at different times of the day. As their condition deteriorates and their visual acuity becomes further weakened, they will reach the level where they are capable only of distinguishing lights, judging whether hands and other objects are moving in front of their eyes, and counting the number of fingers shown in front of their eyes.

(2) Visual field defect

Visual field defect is divided broadly into constriction and scotoma. Scotoma

Normal

Scotoma

Constriction of the visual field

Visual field defects

Figure 7-1 Characteristics of how patients with eye diseases see

is the condition whereby the visual field has degenerated and the field is partially diminished. Constriction of the visual field is the condition whereby the visual field has degenerated at first from the peripheral parts of the view towards the centre and gradually the field becomes narrowed. Visual field defects are the condition in which the visual field is partially lost. The symptoms vary, depending on the causes of visual disabilities, and the symptoms include constriction of the peripheral field (including retinitis pigmentosa), scotoma and defects in the centre of the visual field (including macular degeneration), and partial defects of the visual field (including amotio retinae, glaucoma).

(3) Defective colour vision

Human eyes have cells to detect the three colours red, green and blue, and defective colour vision is the condition whereby one of the three basic colour mechanisms is absent or not functioning. Those persons with defective colour vision have a difficulty in discriminating combinations of colours, and in particular the colours related to red and green are difficult to discriminate. Some of these persons see dark-red or dark-brown colours as black colour, while some cannot distinguish pink and grey. At stations, for example, if the colour of electric billboards is black, persons with defective colour vision cannot see the letters in red at all and they also cannot distinguish letters amongst red and green. Persons that suffer from cataracts caused by ageing have irregularities in their crystalline lens, and they often cannot distinguish among colours related to yellow and blue. Allegedly, more than 10 % of people suffer from some sort of defective colour vision in Japan.

Ageing-related visual disabilities

Parts of the body change as people become older, and the eyes are not an exception. According to a fact-finding investigation on persons with physical disabilities which was conducted by the Ministry of Health, Labour and Welfare in 2006, the population of persons with visual impairment of the age of 20 and over is 310,000, and 49.4 % of the population with such impairments are people of the age of 70 and over (44). This result suggests that developments in medical fields have succeeded in reducing the number of blind people among young people. However, an increase is also seen in the number of cases of age-related impairments, including decreased vision and construction of the visual field.

For example, the most common cause of acquired and developmental blindness is retinopathy of diabetes, and the number is high among elderly citizens. Other causes of visual impairments among elderly citizens are as follows:

(1) Cataract senilis

The symptom often starts from the peripheral parts of the view and it progresses towards the centre of the view. The symptom is difficult to notice at the beginning and the speed of progression varies significantly among individuals. It is not rare that symptoms do not show up until death. Those persons who suffer from cataract senilis see lights scattered, and they have difficulty in seeing things even in bright circumstances and feel that the objects they see are bright, as if they are looking directly at an illuminant.

As a result, they feel that their eyes are very tired and they also feel pain at the back of the eyes. If the symptoms progress further, the iris of the eye become clouded in a whitish colour and they lose sight eventually.

(2) Retinopathy of diabetes

Retinopathy of diabetes is one of the diabetic triopathy. A disorder of the carbohydrate metabolism causes irregularities to body parts, including the retinas of the eyes of people with diabetes. This causes a deterioration of visual acuity. This is the second biggest cause of the loss of eyesight after birth in Japan. Diabetes commonly affects neuropathy, retinopathy and nephropathy, and these three are called the diabetic triopathy. The person notices no symptoms at all at the beginning. Gradually, they suffer from fundal haemorrhage or macular edema, and their visual acuity deteriorates. They also suffer from metamorphopsia. If the person suffers from an intravitreous haemorrhage or fundal haemorrhage in a broad area, they also suffer from loaters and a sudden decrease in visual acuity. When people suffer from neovascular glaucoma, it may cause eye pain, irreversible blindness and phthisis of the eyeball.

(3) Age-related macular degeneration

This is a disorder where ageing causes degeneration of the macular area on the eye retina. This can cause blindness. This used be called senile disciform macular degeneration, or ARMD. A detailed description of this disorder can be found on the Internet site of the Japan Intractable Diseases Information Centre. The initial symptom of this sickness is metamorphopsia, and often people are diagnosed with age-related macular degeneration after seeing a doctor for metamorphopsia. As the symptom progresses, they see things distorted. As they suffer from fundal haemorrhage, their visual acuity deteriorates. They also suffer from central scotoma and sometimes they go blind.

(4) Glaucoma

The optic nerves behind the eyeball are very sensitive and when the intraocular pressure increases, the ball becomes damaged. Once it is damaged, it will never go back to a normal condition. Gradually, the number of the optic nerves decreases and the field of view gets smaller. Glaucoma has become the number one cause of blindness recently in Japan. Constriction of the

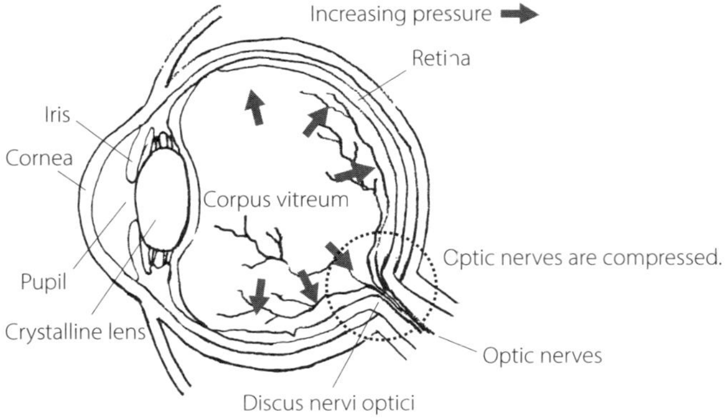

Figure 7-2 Causes of glaucoma

visual field can lead to a terminal condition without it being realised. Regular check-ups are recommended. Once it was believed that high intraocular pressure was the cause for glaucoma. However, it has been found that many people who have normal intraocular pressure also suffer from glaucoma. Today, it is believed that the cause of glaucoma is the brittleness of the discus nervi optici. Intraocular pressure is the biggest risk factor of glaucoma progressing, and the basic treatment for glaucoma is to reduce the pressure in order to stop the progress of the visual field defect. It is reported that by reducing 30 % of the intraocular pressure, the progress of visual field defect stopped for 80 % of the patients with normal tension glaucoma. Glaucoma is found among people who are over 40 years old and who have severe myopia, severe hyperopia or who have relatives who suffer from glaucoma.

Classifications of visual impairment, and the walking of persons with visual impairment

Persons with 1st-class visual impairment are not able to walk by relying on their vision. They need to have a white cane, a walking stick designed especially for visually-impaired persons, or other kinds of assistance. Even among persons with 1st-class visual impairment, some can still distinguish lights or hand and finger movements in front of their eyes. Their conditions in their daily routine are significantly different from those persons with complete blindness. Among persons with 2nd-class visual impairment, those persons with impaired eyesight or visual field defect can walk in familiar places without using a white cane. Persons with 3rd-and 4th-class visual impairment can walk relatively freely. Among those persons with 5th-and 6th-class visual impairment, those persons with impaired eyesight or visual field defects can walk almost similarly to people with normal vision.

As for reading, those persons with 1st- and 2nd-class visual impairment use Braille, and those persons with 3rd- and 4th-class impairment use support equipment when they read writing with small letters. Those persons with 5th- and 6th-class impairment can read writing nearby with glasses, but they have difficulties with reading things such as timetables or the schedule of fees in a station.

The timing of occurrence of visual impairments

When and how impairment occurs can have different impacts on the psychology of different individuals as well as on their daily lives. The timing of the occurrence of impairments is divided basically into two categories: the congenital disorders which occur at birth; and the acquired and developmental disabilities which occur after birth (e.g. in childhood, late-middle age and old age).

Generally, completely-blind persons with congenital visual impairment cannot have images of objects which they cannot touch. However, they often attend and receive training at schools for the blind for such activities as using Braille and walking, hence they often can handle daily living better than persons with acquired and developmental visual impairment.

As for persons whose visual ability is impaired in the later years of their lives, including in their late-middle age and old age, they can receive images of objects or things by using their memories formed before the time of occurrence of impairment. Those persons whose visual impairment developed at older age tend to have difficulties in learning Braille.

People's understanding and thinking of matters or things in daily life can be affected greatly by the timing of the occurrence of impairments. Losing sight suddenly and completely because of an accident, for example, causes a sudden change in living circumstances which comes as a great psychological shock to the sufferer. Those people who have lost their visual sight suddenly and completely often have difficulties in accepting their disability and overcoming their conflict and anxiety.

CHAPTER EIGHT

Transport of persons with visual impairment and their daily lives

Walking with a guide

Walking with a sighted-person is the safest and most efficient way of walking for persons with visual impairment. According to a study conducted by Ichiro Tanaka et al. [45] on the tracks made by people walking with visual impairment, visually-impaired people have difficulties in walking straight and they tend to move in a rightwards direction while walking. When visually-impaired people visit places for the first time, they feel more comfortable and safe if they are guided by sighted-persons. However, this doesn't mean that the mere presence of a guide makes them feel comfortable. It is important for the people who try to offer help to make efforts at reducing any psychological stress held by the persons with visual impairment.

As shown in Figure 8-1, the basic way of providing support to persons with visual impairment is to lead them by letting the visually impaired person hold one's arm (the back of the elbow). When the person with visual impairment wants to use a white cane, they can hold the guide's right-arm and then walk a half or one step behind the guide. This method helps the person with visual impairment follow the movements of the guide, and it helps them know the circumstances around them and the direction in advance every time they both make another step forward. This method also helps them to let their hand go if they need to. The guide must not hold the arm or the white cane of the person,

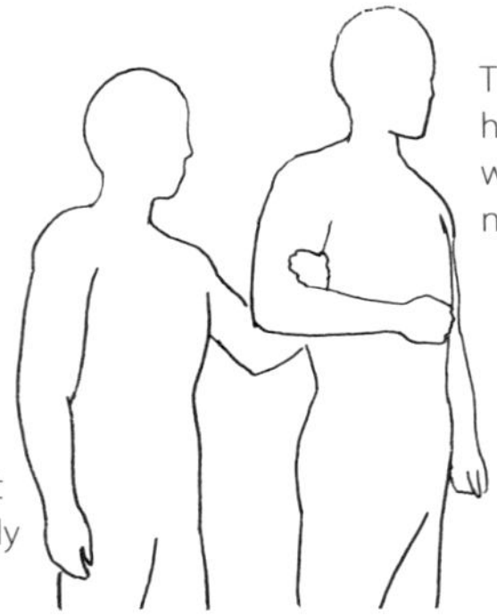

Figure 8-1 The basic way of leading people with visual impairment to necessary places and activities

since the action may make them feel anxiety.

It is important for guides to receive training on how to support a person with visual impairment, and it is equally important for the person with visual impairment to be trained by experts on how to walk. The number of walking-trainers is few in Japan. In 2002, the number of people with visual impairment was approximately 315,000 and the number of the trainers was only 612. These figures indicate that more trainers are needed urgently.

Walking alone by using a white cane

Persons with visual impairment can choose to walk alone only with the help of a white cane. While walking alone by using a white cane, they need to combine their acoustic sense, their sense of smell, sense of touch, cutaneous sense and other senses as well. There are three basic roles in which a white can is used.

One of the roles is to collect information. Users tap the road surface constantly with their white cane while walking, and they discern obstacles and the nature of the road surface through the sounds and touch felt by tapping. By doing this, they can check where they should put their feet a little in advance. The second role is to assure safety. As users tap the road surface constantly with a white cane, the cane touches any obstacles in front of them. This can prevent them from crashing into these obstacles. Sometimes, the cane can act as a cushion between the person and the obstacle and it can ease the shock. But while the cane is suitable for discerning conditions on and near the road surface, it is difficult to prevent the person using it from walking into, for example, side mirrors of trucks and other obstacles which are located high off the ground. The third role of a white cane is to send signals to other people. If a person is walking with a white cane, it can send a signal to other people that he/she has a visual impairment. Such a signal is very useful in the situations like busy streets where cars, motorbikes and other vehicles are driven. The drivers can avoid getting too close and they can communicate to the person in other ways, such as using the horn when it is necessary. Training by experts is required for persons with visual impairment when they use a white cane.

Incidentally, among the total number of people who use a white cane on a daily basis in Japan, 30 % of them are persons with complete blindness. Some people seem to walk with a white cane without any special problems, but these people often have some visibility.

Walking with a guide dog

Walking is one of the most important and basic activities in people's daily life, and using guide dogs is one of the most effective solutions for the difficulties in mobility suffered by persons with visual impairment. Guide dogs are specially trained and they can lead visually-impaired people while avoiding obstacles. This helps people to have independent mobility. When they are walking together, the guide dogs lead the visually impaired persons in certain directions by taking orders from them while securing safety. The guide dogs are specially trained so that they can provide the owners with information about things such as obstacles ahead, intersections, steps, the entrance and exit of buildings and so on.

The users of guide dogs also require special training. For example, when a person with visual impairment wishes to have a guide dog in Japan, he/she needs to spend some time together with the dog and to have special training in order to acquire basic techniques for walking with the guide dog at a training centre. The training usually takes about three weeks. After the training at the centre, the user has two-weeks of walking training around his/her house with special trainers who visit his/her home every day. In Japan, even the dogs which have received special training are not admitted as guide dogs unless a harness is put on, and the user is also required to have a permit on him/her while walking with a guide dog. According to a survey conducted in 2005 by the Nippon Foundation, the number of the guide-dog users was 957. The number of people who wished to have a guide dog immediately was about 4,700, while the number of people who were thinking of having a guide dog in the future was about 3,100. This means that a total of 7,800 people were waiting to have a guide dog [(46)].

Burdens received by visually-impaired people while walking alone

When visually-impaired people who have already received some sort of training from special schools or facilities wish to walk from one place to another, they have the choices of walking with a white cane, a guide dog, a helper or assist tools. However, even when visually-impaired people walk while receiving assistance from such as a guide dog or other assist tools, their psychological stress is much higher than the stress felt by sighted-people when they walk alone. Visually-impaired people may be able to walk relatively freely in places that they are familiar with if they are with some kind of assistance, but a sudden change of routine or unexpected events while walking will cause confusion and psychological chaos to them. There are limits to memorising every detail of the streets and places where they walk around regularly, and also sudden changes or some irregularities are unavoidable. According to a study conducted by Motohiro Ohkura [(47)] on obstacle avoidance, when persons with visual impairment are walking without any orientation hint of the direction, the attention they pay is mostly to keeping in the correct direction. Furthermore, the heart rate of persons with visual impairment becomes much higher when they are walking alone than when they walk with the assistance of a sighted-person, according to a study conducted by I. Tanaka [(48)].

Regarding research by Ikeda et al. [(49)], in a comparison of the heart rates

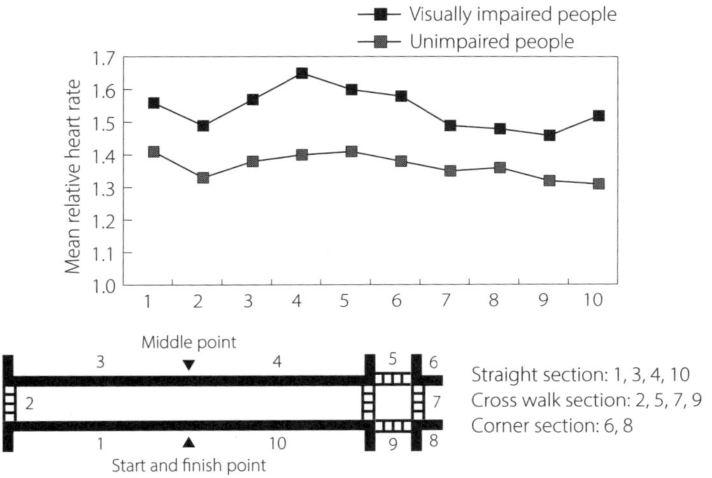

Figure 8-2 Mean relative heart rates

between visually impaired people and unimpaired people, it was shown that visually impaired people's heart rates were higher than unimpaired people's on all sections of the sidewalk (see Figure 8-2). Walking for visually impaired people required not only muscle activity, but also additional mental strain from determining the road-surface condition with a white cane.

Symbol marks for persons with visual impairment

Figure 8-3 shows the universal symbol for visually-impaired people which was enacted by the World Blind Union in 1984. The symbol is used for buildings, equipment and devices which are designed barrier-free in consideration of the safety of persons with visual impairment. The symbol is also seen on traffic signals, international Cecogrammes and in books.

Shown in Figure 8-4 is a "guide-dog mark" which was designed to raise the awareness of guide dogs in the company of visually-impaired people in Japan. The Law Concerning Assistance Dogs for the Disabled is a law which imposes on public and private facilities the duty to accept guide dogs, partner dogs

Figure 8-3 The universal symbol for visually-impaired people

Figure 8-4 A guide-dog mark

and hearing dogs in the company of the disabled. The law was implemented in public sectors and private sectors in May, 2002, and in October, 2003, respectively, in Japan. In present-day Japan, assistance dogs are permitted to enter department stores, supermarkets, hotels, restaurants and other private facilities as well as public facilities including transportation systems, in the company of persons with physical disabilities. Assistance dogs play their roles not as a pet but as a part of the body of physically-challenged people. The dogs are trained and have acquired social manners, and they are to be kept under good hygiene control. Unfortunately, it is still not rare for persons with visual impairment to be rejected from entering and staying at hotels or restaurants if they are with a guide dog, due to the lack of understanding of guide dogs among ordinary people.

Braille block

The Braille block was invented by two Japanese people, Seiichi Miyake and Hideyuki Iwahashi, in 1965. One day they realised that the boundary between moss and soil could be felt and distinguished by feet even if wearing shoes, and this became a hint for designing the Braille block [50]. The first Braille block in the world was placed on one of Japan's national roads, Route 2, in front of Okayama-prefectural school for the blind in 1967. This became possible after repeated negotiations with the Ministry of Works, the predecessor of today's Ministry of Land, Infrastructure, Transport and Tourism. Following the implementations of the Heart Building law in 1994 and the Barrier-free Transportation Act in 2000, Braille blocks were set in traveller-facilities and on roads, stations, intersections and their surrounding areas as part of creating barrier-free zones. The Braille block shown in Figure 8-5 (b) is the type in which only raised parts of tactile surface of the block are laid on the carpet. The Braille block is blending right in with the rest of the area without losing its effect.

(a) On a station platform (b) In a facility building

Figure 8-5 Braille block

Different kinds of Braille block have been laid in different places for many years, but "Dimensions and patterns of raised parts of tactile ground surface indicators for blind persons" was standardised by JIS in September 2001, following an appeal made by people with visual impairment for standardisation of the

Braille block. Guidelines for the tactile ground surface indicators on platforms of railway stations were implemented in December 2002.

Braille blocks can be divided broadly into three patterns of line-Braille block, dot-Braille block and warning-Braille block for platform-ends, as shown in Figure 8-6. The line-Braille block is used as an indicator for the walking direction. It is usually laid on the ground surface to lead users to a station, facilities and other places. The dot-Braille block is used to draw caution from users. For example, it is employed to tell users that it is dangerous to go beyond the line of the dots. The dot-Braille block is usually laid on the ground surface of places such as the head of stairs, ends of platforms, the entry of intersections, gaps between the sections of line-Braille blocks and stopping places. The warning-Braille block for platform-ends is used as an alarm at platform-ends. This block is made with the combination of dot-Braille block and one line-Braille block, as shown in Figure 8-6 (c). However, only a few Braille blocks have been set as it has not yet been long since the guideline was implemented. Moreover, in Japan, bicycles and motorbikes are often parked in the way and objects such as boxes are placed on Braille blocks laid on sidewalks, and this behaviour by people prevents visually-impaired people from using the blocks properly. This is one of many social problems confronted by physically-impaired people in Japan.

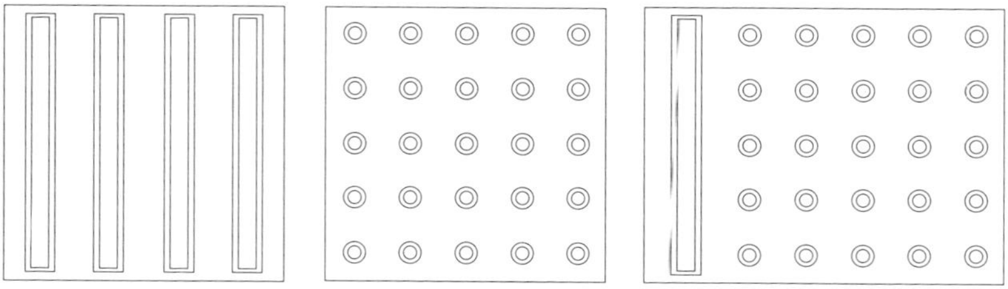

(a) Line-Braille block (b) Dot-Braille block (c) Warning-Braille block for platform-ends

Figure 8-6 Types of Braille block

An additional problem is that there are some cases of the Braille block being stopped by a manhole cover on the sidewalk, and also there is another case in which the path of the Braille blocks avoids the manhole cover unnaturally. In

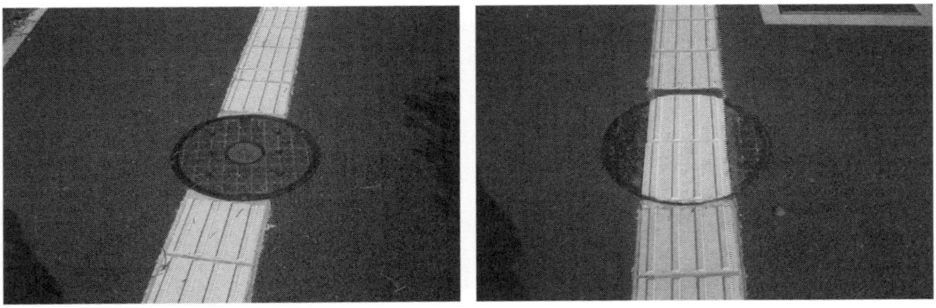

(a) Braille blocks path cut off by manhole (b) Braille blocks painted on a manhole

Figure 8-7 Manhole cover and Braille blocks

such a case, Figure 7 (b) shows a way of solving the problem of Braille blocks on a manhole cover.

Valid width of sidewalks

There are places where the width of side-walks is not enough because electric poles, bus stops, planting zones, and other structural objects are built nearby. It is necessary to keep a valid width on each sidewalk by broadening side-walks and eliminating or shifting electric poles and planting zones. Moreover, in towns, illegally-parked bicycles and shops' display name-boards are often

Figure 8-8 Illegally parked bicycles and shop displays near Braille block on a side-walk

placed on sidewalks and they are crowding the space on the walks, as shown in Figure 8-8. According to research on the walking behaviour of persons with visual impairment conducted by the author, such persons often need to stop on sidewalks in order to check the surface of Braille block or roads every time when their white canes touch the covers of side ditches. They also need to stop occasionally when they encounter obstacles placed by people on sidewalks. Illegal parking of bicycles, the placing of shop displays and some other moral-related problems should be solved by teaching children that such behaviour causes trouble for people with visual impairment, people in wheelchairs and other physically challenged people.

The difference in levels on pedestrian crossings

The difference in levels between roads and pedestrian crossings is built purposefully so that people with a white cane can detect the boundary easily. The difference is set at 2 cm by the Ministry of Land, Infrastructure, Transport and Tourism in Japan. The height of difference was set by the Ministry after conducting experiments involving persons with a physical disability in wheelchairs and visually-impaired people. However, the difference in the level off the ground is too high for elderly people in wheelchairs to climb up as the strength in their arms tends to be weak, while persons with visual impairment also cannot always feel the difference correctly. Incidentally, the difference in level is sometimes too high for even unimpaired people, according to a survey conducted

Figure 8-9 The difference in level on pedestrian crossings

by Hiroshi Ogami et al. [(51)]. For example, users of baby strollers which have small wheels need to pick up the front and rear wheels alternatively in order to go over the difference in height levels.

White canes

White canes are a white walking stick used by persons with visual impairment when they walk. Generally, the diameter of the cane is 2 cm and the length is around 1.0-1.4 m. A white cane acts mainly for information gathering, such as learning the difference in the level on roads and the gaps among roads, as well as acting as a system to remind other vehicle drivers and pedestrians about the presence of a visually-impaired person on roads. A white cane is titled as a safety cane for the blind in the Physically Disabled Persons Welfare Act and is included in the category of welfare tools in Japan.

It has been a long time since walking canes became an essential tool for persons with visual impairment, and it was after the First World War that walking canes were painted white and made shiny. It is said that an English person called James Biggs first painted the cane white so that the users could be noticed easily by other people even on busy streets. It is also said that a French person, Guilly d'Herbemont, designed the current shape of the white cane after she saw visually-challenged people were in danger as the number of vehicles increased on roads. She got a hint from the baton which her husband was using. She also encouraged unimpaired people not to use white-coloured walking sticks.

The currently used white canes are divided mainly into a standard type and a portable type. The standard type has no joints, and hence it has excellent transmissibility. This type suits those situations when the users walk alone. The fold-type and slide-type have less transmissibility than the standard type, but they are portable and convenient for carrying when the visually impaired are using a bus, train or some other transportation system. Some modern white canes are embedded with an ultrasonic sensor and they can convey the existence of obstacles to the users by vibrating.

The canes are usually in white, but yellow-coloured canes are also available.

Only white-coloured canes were recorded in the Road Traffic Law in 1960 in Japan. However, when the Law was revised in 1971, the yellow-coloured canes were added to it. The reasons for adding yellow-colour canes are that yellow is often easier to notice than white and white-colour assimilates into the background in places with snow. The difference in colours is not used to identify the differences in types and severity of disabilities among people.

Paying attention to people walking with a white cane

People walking with a white cane are not necessarily in need of immediate help. When a person is walking with a white cane, people around the person should pay attention to see if he/she is at great risk. When, or if, a person with a white cane is at great risk of, for example, falling from a platform or steps, and when, or if, the person looks as if he/she is in trouble, support given by the people around the person would be helpful.

When visually-impaired people walk with a white cane, they usually focus their attention on the tip of the cane. Guiding them by holding the hand with the white cane or holding the cane itself prevents the person from gathering information. People should give help to them by standing on the opposite side from the hand holding the cane. Persons with visual impairment have difficulties in grasping what other people are trying to do with them since they cannot see the situation with their eyes. Guiding them by touching their shoulder or hand is often felt, by visually impaired people, as if they are being pushed from the back—hence, they feel anxiety. Verbal communication while guiding to them is a good way to make them comfortable, and clear indications such as left or right are more helpful than vague words such as there or over there.

In present-day Japan, as a part of creating barrier-free environments, special training and education for helping the disabled are given to the staff and personnel in some facilities and public places, since they are likely to encounter the physically-challenged people.

History of guide dogs

In 1819, in Vienna, a priest named Johann Wilhelm Klein trained a dog so that the dog could help persons with visual impairment. The dog trained by Klein is thought to be the first guide dog in the world. The first facility for training dogs as guide dogs and as a part of welfare services was established in Germany after the First World War. The purpose of this facility was to provide support to military men who had had visual impairment caused by the war.

The first four guide dogs were imported from Germany to Japan in 1939. The history of training of dogs as guide dogs in Japan began in 1957, and guide dogs have been often called "eye-mates" in Japan after 1972. The Japanese Road Traffic Law was revised in 1978 and it was stipulated that persons with visual impairment must walk with either a white or yellow cane or a guide dog which is specified by Cabinet Order. The entry of guide dogs to public transportation systems as a companion of visually-impaired persons was granted in accordance with the revision of the law. In 2002, the Law Concerning Assistance Dogs for the Disabled was enacted. This law works to impose on public facilities the duty to accept guide dogs, partner dogs and hearing dogs in the company of the disabled. The law bound only public facilities to any performance of the obligation at the beginning, but in 2003, restaurants, hotels and other private facilities began accepting the entry of guide dogs.

Guide dogs are given training at facilities appointed by the National Public

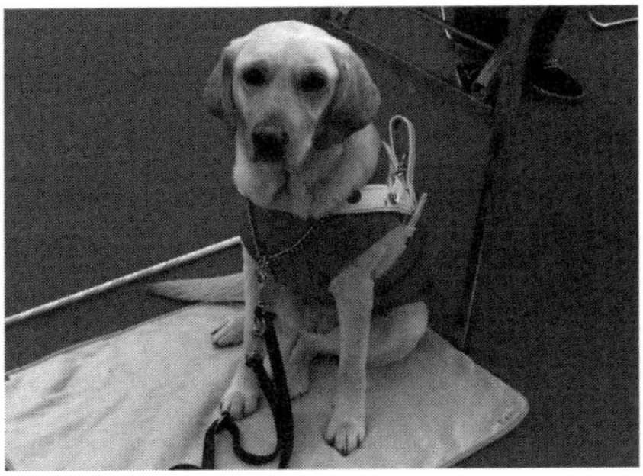

Figure 8-10 A guide dog

Safety Commission in Japan, and it takes approximately one year to complete the training. Currently, there are nine facilities for training in Japan. Labrador Retrievers and Golden Retrievers are the most common types of guide dogs in Japan, and the total number of guide dogs was 1,043 in 2012 according to the Ministry of Health, Labour and Welfare.

Traffic-light machine with an acoustic device

A traffic-light machine with an acoustic device is a traffic-light machine designed for visually-impaired people. The traffic light with an acoustic device can inform the people of the change of lights by sounds. The Japanese first traffic-light machine with an acoustic device was introduced in Tokyo in 1955, and it informed people of the change of lights by producing the sound of a ringing bell. Since then, the machines have been put in places that visually-impaired people are likely to use frequently, and the number of the machines accounted for in 2006 was approximately 14,200 (including 12,750 of imitative-sound and 1,450 melody-sound). Some of the machines have a speaker embedded as shown in Figure 8-11. The colour of the speakers used to be yellow, but white or gray coloured speakers have become more common.

In present-day Japan, there are two types of traffic-light machines with an acoustic device available. One is a normally-on type, which produces sound every time when the light changes to green, and the other is a push-button

Figure 8-11 The push-button of a traffic-light machine with acoustic device

type, which is activated only when a button is pushed. The push-button style is more common, except in some parts of city centres. This is because many complaints have been made about the noise frequently made by the device of the normally-on type. Today, the sound produced by the normally-on type is often controlled so that no sound is produced at night.

An imitative-sound and a melody-sound were mainly used as guiding sounds originally and the sound of a bird singing is applied to both the imitative and melody types. The imitative-sound style of traffic-light machine with an acoustic device employs a system whereby different sounds are put out alternatively from the two speakers located at each side of a pedestrian crossing. Both the imitative- and melody-styles were, however, often used in the same area, and this confused users. In 2003, Japan's National Police Agency issued an official notice on improving the conditions [(52)], and one of the imitative-sound types with alternative-sound production has gradually disappeared. Since then, studies on traffic-light machines with an acoustic device have been conducted in cooperation with the Japan Federation of the Blind, and it has been proved that the imitative-sound type with alternative-sound production has a high induction property. As a result, the National Police Agency has established installation and investment policies, and it has been decided to standardise the use of the imitative-sound type with alternative-sound production throughout the country.

CHAPTER NINE

Public transportation systems in Japan

Getting-on and getting-off trains, and the reserved space for persons with physical disabilities

Shown in Figure 9-1 is a wide ticket-gate (90 cm-width) which is designed for wheelchair occupants to go through in a wheelchair. In Japan, most stations have at least one ticket-gate with this similar width, and the wide gate is usually located at the very end of the line of gates. However, the size is not quite wide enough for a wheelchair to go through comfortably, and thus the occupant needs caution while propelling the chair.

A call-button is usually set up on each platform of stations (see Figure 9-2).

Figure 9-1 A wide ticket-gate

Any person with physical disabilities can push the call-button and have verbal communication with station attendants. When the disabled person needs assistance, a station attendant comes and waits for the arrival of a train together with him/her, and they provide assistance when the train has arrived. If a wheelchair user needs assistance, a station attendant sets a folding board across the gap between train and platform after the door of the train has opened. The wheelchair occupant can get on the train while sitting in the chair. While the occupant is moving into the train, an announcement is being made to say something to the effect of the event and the train will not start moving until the action of transferring the wheelchair is completed. The staff at the starting station inform the station where the occupant will get off that a wheelchair user is on the train. The occupant can thus receive the same assistance at the destination. When the wheelchair user is coming down the board from the train, it is better to come down backwards in order to avoid being thrown out from the chair in the case of the chair tipping over.

Figure 9-2 A call-button on a platform

Figure 9-3 shows a reserved space for wheelchair users in trains. Installation of the reserved space is obligated by the Barrier-free Transportation Act. The reserved space is often installed in new cars of trains, and each space is usually equipped with an emergency call-unit with which the users of the space can communicate verbally with the crew of the train. The space is also equipped with tools which can be used to secure wheelchairs to the train. The reserved

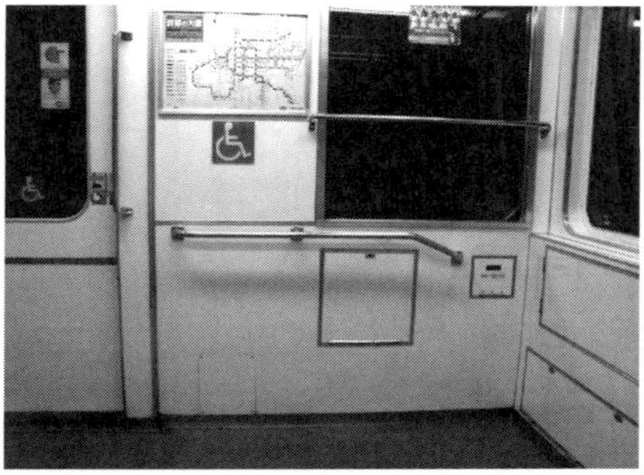

Figure 9-3 The reserved space for wheelchairs in a train

space is created at the end of a carriage and it has arm rails and an intentional mark for persons with disabilities which is placed on the wall.

When persons with visual impairment are getting on or getting off from a train, they need to be careful about the gap between the train and the platform. In the case of getting on a train while receiving assistance from a guide, the visually-impaired person should walk directly towards the platform and wait at the edge of the platform. The guide gets on the train after saying, "I am getting on the train" and the person follows the guide. If the guide gets on the train by following the basic way of leading persons with visual impairment (see Figure 8-1), the visually-impaired person behind the guide may step into the gap between the train and platform. Special caution should be made for avoiding such an accident. When guiding the person to sit on a seat in the train, the guide leads the person so that his/her knees touch the seat first and then move the person to sit down. The guide needs to have constant verbal communication with the visually-impaired person during guidance.

Elevators and escalators at stations

Each elevator at stations should be big enough for a wheelchair to enter and exit the elevator, as well as to turn inside the elevator. Furthermore, the size, shape and colours of control buttons should satisfy a certain criteria and they have to be user-friendly for visually-impaired persons. Elevators are usually

Figure 9-4 The mirror and special control board inside an elevator

equipped with a special control-board for wheelchair users. The board is placed at a low position on the wall of the elevators. When the wheelchair users press buttons on the special board, the door of the elevator opens longer than usual so that the users can make entry and exit comfortably. A mirror is installed inside some elevators. The mirror is used by the wheelchair users for checking whether the door is open or closed, or whether other people are getting on or getting off. To create an ideal elevator for wheelchair users, the width of the entry and exit should be 80-90 cm or more, the height of arm rails should be 80-85 cm from the ground, and the shape of the rail should be round.

Each escalator at train stations is equipped with a call-button which is used for calling for assistance (see Figure 9-5). The international mark for persons with disabilities is placed nearby the escalator. When a wheelchair occupant wants to use an escalator at a station, a station attendant first places a tape across the landing of the escalator in order to control the entry of other people temporarily, and then the attendant stops the escalator. Three steps of the escalator become flat and the attendant assists the occupant of the wheelchair to move onto the flat space. A stopper which is used to prevent the wheelchair from moving will be set, and the wheelchair occupant puts on the brake to secure the chair. Usually, each step of the escalator is more than 100 cm-width, and it is rimmed with a colour which is easy to notice. If escalators are not designed for use for wheelchair users, helpers or another form of help are necessary.

Figure 9-5 A call-button placed at the entry of escalators

Barrier-free buses

The floor of buses is usually high off the ground in Japan and the steps are placed at the entry/exit door. These structures of steps have been a big obstacle for elderly citizens and persons with physical disabilities. As a part of the creation of barrier-free zones, improvements have been made on steps, floor and other parts of buses. Today, the floors of many buses are low off the ground, and buses have been remodelled and some of the remodelled buses are installed with a lift. Moreover, some buses have been remodelled so that the buses have one or no step at the door. In particular, the buses with a lift have had a great effect in increasing the number of bus users among wheelchair users.

(1) Non-step bus

A non-step model of a bus was first introduced in 1997. This model has no step at the entry/exit door, and the floor is designed to be low off the ground. The height from the ground is approximately 30 cm off the road surface, and the floor level can be adjusted lower when it is necessary. Traditionally, bus users have to climb one or two steps up when they ride on a bus. When getting on a non-step bus, if the user goes one step up, their feet are on the floor of the bus already. As shown in Figure 9-6, a call-button is installed at the entry/exit door of the bus so that the user can communicate with the driver if it is necessary. Almost all non-step buses are installed with

Figure 9-6 The entry of a non-step bus

special devices useful for wheelchair users, and a ramp-board can be put between the bus and the road surface in order for the wheelchair occupant to drive up on a gentle gradient. When the ramp-board is placed between the bus and sidewalk on roads, for example, the gradient is very close to the flat surface. However, there are still some problems remaining for non-step buses. For example, because of the structure of non-step buses, there are differences in the level at the rear part of bus, and the number of seats in the bus is reduced in order to secure space for a wheelchair. Moreover, the cost of the model is expensive, and hence each introduction of the model is not possible without subsidy.

(2) One-step bus

A one-step bus is a model where the steps at the entry and exit have been reduced from two to one. The level of the floor of the bus is approximately 50 cm off the road surface. Though it still has a step at the door, the level of the floor is generally low and this makes it easier to get on and off the bus. If a ramp-board is used when it is necessary, wheelchair occupants, especially electric-chair users, are able to get up the ramp into the bus without any help. The main differences between non-step buses and one-step buses are the inner structure and the cost of the bus itself. In addition, there is almost no difference in level inside of a one-step bus and there is no wasted space either. Therefore, it can ensure that a good number of seats are available for other bus users.

(3) Buses with a lift

With the traditional model of bus, which has two-steps at the entry/exit door, getting on and off with a wheelchair is difficult and it requires having people to carry the wheelchair in and out of the bus. As part of a promotion for social participation by wheelchair users, changes were to be made to buses—traditionally the closest transportation system for people. As a first step in the change, buses with a lift were introduced. Each bus with a lift costs approximately 10 million Japanese yen, and it requires the bus to stop for a longer period of time when the lift is used for a wheelchair and this affects the schedules of buses. Buses with a lift gained some popularity once, but since one-step buses and non-step buses have been introduced, the popularity of buses with a lift has declined. In present-day Japan, the number of tourist buses with a lift is increasing and more buses are built with a low-level floor.

Demand-responsive buses and special transportation service

Demand-responsive buses are a bus service to provide bus service flexibly in accordance with the demands from the users, and the buses in this service do not follow the normal bus schedule. This service belongs to the public transportation system and its role is somewhere between the main bus-service and taxi services. A demand-responsive bus-service was first introduced in Japan in 1972. One of the advantages of this service is that the buses are not required to stop at each bus stop and they do not need to stop when no customer is waiting. One of the disadvantages of this service is that it is difficult for the passengers to estimate how long it takes to their destinations. The cost of running the service is more expensive than the regular bus service.

A special transportation service (STS) is a door-to-door transport-service which is usually used as a daily transport-system for elderly citizens, persons with disabilities and other people who have difficulties. The vehicles used for this service are usually equipped with a lift and have care-helpers. This service was introduced in 2000 following the relaxation of market-entry regulations for buses and taxies. It began as a part of the revision of the Road Transportation

Act. However, problems related to its management control and accidents have loomed since the introduction, and it requires further improvement in safety arrangements.

There are three types of services in the STS. One of the services is called the door-to-door service and the targets of this service are people who are not able to use, for example taxis, alone. The vehicles used for this type of service are equipped with a lift and are remodelled as barrier-free vehicles. The operation of the service is often conducted by local taxi companies or welfare organisations which have received a commission for its operation from the local government. The operations are thus usually conducted while receiving a subsidy from the local government. The second service is called a scheduled service. This service is similar to the standard share-ride buses. The intervals between stops are shorter than a standard bus service, and the vehicles used for this service are also equipped with a lift and the floor is built low. The third service is a service where vehicles make round trips among public facilities. The service is generally initiated by local government, and the vehicles are remodelled as barrier-free vehicles and stop at hospitals and public facilities.

Such services make it possible to provide users with more complex services than the conventional public transport system. These services afford an opportunity for people with physical disabilities to use the services independently as useful public transportation.

Light-rail transit

Light-rail transit (LRT) is a medium-size city-transportation system. Under this system, a few relatively small-sized cars run on rails which are laid on roads. The term LRT was first used in America in order to distinguish it from the street-car and tramway. The term was also effective in creating a new image for such a service. LRT has become a useful and convenient transport system for people who live in cities, and it has played an effective role in improving the convenience and activities of disabled people.

Many street-car and tramway services were abolished because the punctuality of the services deteriorated as many other vehicles drove on the rail areas of these services. In recent years, there are many cases where the entry of other vehicles is restricted, and other regulations have been established in order to keep vehicles off the rail areas of the service. Naturally, it is difficult to obtain enough space for running transport systems like LRT in city centres. In such situations, it is possible to build grade-separated crossings and an underground. In fact, a part of LRT was built underground in the 1960s in Germany.

In Japan, tramway and LRT services have been re-examined gradually as a countermeasure for reducing traffic quantity and the number of traffic jams. Akihiro Mihoshi et al. [53, 54] pointed out that introducing LRT as a public transportation system could solve traffic-related problems, and also it is related to the enlivenment of communities and towns.

As a way of creating a barrier-free environment, the floor-level of the cars used for a LRT system is usually low, and the platforms of stops are built high off the ground. The low floor and high platforms can solve problems which wheelchair occupants confront when they get on and off the cars. Moreover, the stops are built nearby other public transportation systems in order to create a seamless traffic system. For instance, building an LRT stop and a bus stop together, or extending the stops on the roads to a station square, makes changing cars and trains become much smoother. By doing this, the convenience of the public transportation system is improved significantly.

Figure 9-7 shows an old type of tramway of a Japanese private railway. As with the modernization of the city, the surrounding area of buildings was con-

Figure 9-7 An old type of tramway in Japan

verted consequently. However, this modernization was not planned to include this tramway, therefore, the tramway station is left in the middle of the traffic. Not only a handicapped person, but even a physically unimpaired person has trouble to use the station.

Park and ride system

Park and ride (P&R) is a system under which people commute to schools or companies by taking their own vehicles to stations nearby and then changing to LRT at the stations. This system is not popular yet in Japan in comparison to European countries. In order to promote the P&R system, some local governments and railway companies set up a discount system for parking vehicles. One of the reasons why this system has not been accepted widely yet in Japan is that large-size parking spaces are necessary for establishing the P&R system, and the land price in cities has become an obstacle for its promotion. The P&R system is effective in cities where the public transportation system is well developed, but it is not very effective in cities where there are not many bus and train services which are available, or where cars and other vehicles are daily essentials.

Previously, almost all car parking costs were calculated per hour. However, recently, it is possible to pay for a fixed limit of half a day or a daily unit at a cheaper cost, so it is easer to use the car park. Figure 9-8 shows an unmanned

Figure 9-8 Unmanned car park

car park which is very common. Each parking space is numbered, and when leaving the car park, the user has to push their own number and pay at the machine.

Barrier-free airplanes

When wheelchair users use airplanes, they must get on airplanes before other passengers get on and they must get off the plane after other passengers get off, for safety reasons. Because of this, it is necessary for wheelchair users to check in early. Unlike getting on vehicles such as a bus, train and other modes of public transportation, when wheelchair users get on planes they usually have to switch to an in-flight wheelchair from their own wheelchairs. When the wheelchair user employs a chair of the basic type, he/she needs to switch to an in-flight wheelchair and book his/her chair as luggage. When the wheelchair is a bed-with-wheels or some other special type, the occupant can be taken to the boarding gate. Incidentally, the basic type of chair may be used until just before boarding today.

A variety of types of wheelchairs are prepared by airplane companies, and all of them are the types used for nursing care. Wheelchair users are advised to contact to the airline company which they use in advance. In-flight wheelchairs are often a simple and basic type and they are not comfortable to sit on for a long time. The occupants need to use something such as a cushion on the seat if the flight is going to be a long one.

Subsidy for each public transportation system

If a person has been granted an identification booklet for disabilities and wants to use public transportation, they should show the booklet and then their bus and train fares are discounted by up to 50 % in Japan. Their taxi fare will also be discounted by 10 %, and some municipalities distribute free riding-tickets to people with the booklet. Their fare for boats and other maritime transportation may also be discounted, but the rate varies among the maritime transportation companies. In principle, the discount system for transportation is applied only to disabled passengers, and the fare for cars and other vehicles is not discounted usually. Some prefectural parks and other facilities offer a discount on using parking space for people with the identification booklet. Whether the system is available or not depends on cities and prefectures, and some local governments offer the use of public parking spaces free of charge.

Epilogue

This book has described the problems faced by persons with disabilities and focus is placed especially on the mobility of persons with physical disabilities. If the environment for the mobility of persons with disabilities is improved significantly, their restricted mobility will be eased and their choice of jobs and activities will increase. As a result, they would find greater motivation for fulfilling their lives. Furthermore, if they can go out freely, then they can enjoy the very flavour of living. Improving the user-friendly mobile-environment is thus extremely important for persons with disabilities.

Barrier-free sidewalks, for example, should be created, together with convenience and safety in the entire routine of movement, from the user's starting point to the destination. A partial barrier-free environment is not effective. It is important to improve the environment for movement and assist-devices so that persons with disabilities can manage transportation without special help from care-givers. It is also important to improve road and street conditions so that obstacles for those persons with disabilities can be removed. However, removing obstacles on streets may create other problems. For example, when the rolling phenomenon is eliminated from sidewalks, super-elevation is created on sidewalks, even though it is hard to see with the naked eye. Super-elevation is created in order to let rainfall flow towards the road side. Such sidewalks may appear to be easy to drive on for wheelchair users, but in fact super-elevation causes wheelchairs to move towards the road side against the users' will. Under such circumstances, wheelchair occupants receive a physical burden on their muscles constantly and this will have a negative impact on their physical condition. We should know that creating barrier-free environments do not always reduce burdens on the users equally.

The earth-moving/construction industry in Japan has conducted hearings and questionnaire surveys in order to understand whether the creation of barrier-free environments are actually improving the mobility of disabled users and reducing their physical burden. The survey method can collect a great deal of opinions and knowledge on a topic relatively easily, but the biggest demerit of such a method is that it is not objective. The effects of improving the environment have to be evaluated objectively by employing appropriate barometers. Therefore, it is necessary to develop forms of measurement which indicate the ease or difficulties of mobility for disabled people. Based on the author's experience, various kinds of approaches to evaluation have to made, together with multiple forms of measurement, in order to understand the relation between human beings, machines and the environment.

To improve the environment for mobility, experts in various fields have to cooperate together. When Japanese people create something, they tend to forget about the existence of human beings, and they tend to get lost in whomever they are making the product for. It is important that the ideas of various people, with various points of view, in various kinds of studies, are taken in positively in order to create something precious and permanent.

I would like to thank a number of people who have kindly offered help and advice in preparing this book. I would like to thank also the people who took part in research as the subjects of experiments—without their participation and cooperation, none of the experiments would have been possible. I would like to thank Prof. Akihiro Mihoshi of Kansai University of Welfare Sciences; Prof. Masaru Nakaseko of the Graduate School of Health Care Sciences, Jikei Institute; Prof. Shigeyuki Minami of Osaka City University; Prof. Kazunari Morimoto of the Kyoto Institute of Technology; and Prof. Isao Takeda and Prof. Hiroaki Hirose of Takarazuka University of Medical and Health Care for their advice in the making of this book.

Finally, I would like to thank Ms. Kako Richards for assistance with the translation, and Dr. Ian Richards for his help.

With appreciation to all of the people who helped me, I hope that this book can be somehow useful for those who are going to study the environment for the mobility of persons with disabilities and for those people who are involved in creating and improving such an environment.

This book is to be published with the assistance of a 2012 Grant-in-Aid for Publication of Scientific Research Results from the Japan Society for the Promotion of Science (JSPS).

References

(1) Cabinet Office, Government of Japan (2012). *Annual report on government measures for persons with disabilities.*

(2) Cabinet Office, Government of Japan (2012). *White paper on aging society.*

(3) Wolfensberger, W. (1981). *The principle of normalization in human services*, National Institute on Medical Retardation.

(4) Maslow, A. H. (1966). *The psychology of science: A reconnaissance.* Harper & Row.

(5) Kurokawa, A. (1983). *Fukushi wa ikani arubekika: Shiminfukushi no genjou to kadai* (How should welfare be?: The present conditions and problems of citizen welfare). Seishin Shobo.

(6) Ministry of Health, Labour and Welfare (2009). *Heisei 20 nendo kaigosaabisushisetsu/jigyousho chousa* (2009 investigation of care service institution and establishment).

(7) Care Work Foundation (2010). *Heisei 21 nendo kaigoroudousha no shuugyoujittai to shuugyouishiki chousa* (2009 investigation of actual working situation and awareness for care workers).

(8) Ueda, S. (1996). *Rehabiriteishon: Atarashii ikikata wo tsukuruigaku* (Rehabilitation: Medical way of making a new life). Kodansha.

(9) Ueda, S. (1994). *Medemiru rihabiriteishonigaku* (Pictorial medical rehabilitation). University of Tokyo Press.

(10) Ueda, S. (1980). Shougai no juyou: Sono honshitsu to shodankai ni tsuite (Acceptance of disability: Essence and stages. *General Rehabilitation*, Vol. 8 (7), pp. 515-521.

(11) World Health Organization (1980). *International classification of functioning, disability and health.*

(12) Shimazu, A. (1996). *Roudou no genten "tachii/furumai" no subete: Sagyoushisei/dousa*

no zusetsu (The origin of labour "behaviour": Illustrated work posture and movement). Occupational for Safety and Health Training Institute

(13) Masuzawa, T., and Minami, S. (2010). Current status and the future of electric wheel chairs in Japan. *Journal of Human Environmental Studies*, Vol. 8 (1), pp. 45-53.

(14) Japan External Trade Organization (2000). *Fukushiyougu kyufuseido nimiru oubei no koureishafukushi* (Financial support system of medical aid equipment for the elderly in western countries).

(15) Hettinger, T. (1960). Muskelkraft bei mannern und frauen. *Zentralblatt Arbeit und Wissenschaft*, Vol. 14, pp. 79-84.

(16) Engström, B. (1993). *Ergonomics: Wheelchairs and positioning*. Posturalis Books.

(17) Nagai, S., Mizuno, M., Tanimoto, M., Konishi, N., Kimura, N., Tsukagoshi, T. (2002). Influence of wheelchair adaptability given by the physical therapist's participation in wheelchair prescription: From the wheelchair use state investigation in the hospital. *Hokkaido Journal of Physical Therapy*, Vol. 19, pp. 50-56.

(18) Ministry of Land, Infrastructure, Transport and Tourism (2003). *Douro no idou enkatsuka seibi gaidorain* (Guidelines for the maintenance and improvement of easier accessibility and usability of transportation infrastructure).

(19) Ikeda, H. (2010). Evaluation of physical load of hand-rim wheelchair propulsion on barrier-free model courses. *IATSS Research*, Vol. 34, No. 1, pp. 48-54.

(20) National Rehabilitation Center for Persons with Disabilities (1994). *Physically handicapped people/senior citizens and driving: Historic process and present conditions*. Chuohoki Publishers.

(21) National Police Agency (2011). *Driver's license statistics.*

(22) Dols, J. F., Garcia, M., and Sotos, J. J. (1996). Procedure for improving the ergonomic design of driving positions adapted for handicapped people. *Boletín Factores Humanos*, Vol. 12 (13), pp. 7-17.

(23) Kember, P. (1992). Vehicle control tests for the disabled could influence ergonomic design. *Automotive Engineer*, Vol. 17 (5), pp. 52-55.

(24) Endo, K. (1993). Shogaishayou jidoushaunten hojosouchi (Driving equipment for handicapped people). *Study Current Rehabilitaiton*, Vol. 77, pp. 24-29.

(25) Wahira, Y. (2002). *Daredemo wakaru koutsu no bariafurii* (Easy to understand barrier-free traffic). Seizando-Shoten Publishing.

(26) Ministry of Land, Infrastructure, Transport and Tourism (2001). *Shogaisha/koureisha no tameno koukyoukoutsuukikan no sharyoutou ni kansuru moderudezain* (Model design about vehicles of a public transportation system for handicapped people and elderly people). Institution for Transport Policy Studies.

(27) Nitta, Y., Odani, M., and Yamanaka, H. (2000). *Machizukuri no tameno koutsuusenryaku* (Transport strategy for inclusive society). Gakugei Shuppansha.

(28) Nippon Foundation (1998). *Idoushudankakuho no tameno 10,000 nin ankeito* (Questionnaire survey of 10,000 people about secure transportation).

(29) Ikeda, H., Kitagawa, H., and Mihoshi, A. (2011). Summary of research on driving support system for upper and lower body impaired people. *Proceedings of 11th European AAATE Conference*, pp. 834-839.

(30) Ikeda, H., and Mihoshi, A. (2011). Wright, A. M., and Rothenberg, S. P. eds. *Posture: Types, assessment and control.* Nova Science Publishers.

(31) Ikeda, H., Mihoshi, A., Kimura, N., and Amemiya, K. (2008). Driving behaviour on curves for wheelchair users. *Traffic Science*, Vol. 39 (1), pp. 60-65.

(32) Ikeda, H., Mihoshi, A., Ikeda, H., and Hisari, Y. (2007). Physical load related to highway-driving among disabled people. *IATSS Research*, Vol. 31 (1), pp. 100-109.

(33) Ikeda, H., Mihoshi, A., and Kimura, N. (2007). Problems of driving vehicles for disabled people. *Journal of Human Environmental Studies*, Vol. 5 (1), pp. 27-33.

(34) Ikeda, H., Hirose, H., and Mihoshi, A. (2010). Study of movements of the head and muscle load of the shoulders and lower arms in driving for wheelchair drivers. *Journal of Human Environmental Studies*, Vol. 8 (1), pp. 75-79.

(35) Ikeda, H., Hirose, H., and Mihoshi, A. (2010). Driving and load on curves of upper and lower body impaired people. *Proceedings of 12th International Conference on Mobility and Transport for Elderly and Disabled Persons*, CD-Rom.

(36) Arai, Y. (1997). *Nou no shikumi* (Structure of the brain). Nippon Gitsugyo Publishing.

(37) Berger, K. W. (1970). *The hearing aids, its operation and development*. The National Hearing Aid Society Detroit.

(38) Boer, B. (1984). Performance of hearing aids from the pre-electric era. *Audio Acoust*, Vol. 23 (2), pp. 34-35.

(39) Kajiya, T. (1990). Hearing aid : Its history and problems about wearing. *BME*, Vol. 4 (1), pp. 2-10.

(40) The National Center of Speech & Hearing Disorders (1986). *Choukakushougaisha anzenuntenyou kigukaihatsukenkyuu ni kansuru houkokusho* (Report concerning research of driving support equipment for hearing impaired people).

(41) Miyazaki, S. and Ishida, A. (1985). Traffic-alarm-sounds monitor for the hearing handicapped drivers. *Japanese Journal of Medical Electronics and Biological Engineering*, Vol. 23 (5), pp. 306-311.

(42) Katoh, H., Kanaya, M., Kim, Y., Yoshioka, K., Itoh, A., Mizuno, M. (2001). Cases of dizziness during driving. *Equilibrium Research*, Vol. 60 (3), pp. 188-193.

(43) Oshima, M. (1986). *Koutsuu to ningen* (Traffic and Human). Nippon Hyoron Sha.

(44) Ministry of Health, Labour and Welfare (2008). *Survey on persons with physical disability.*

(45) Tanaka, I, Shimizu, O., Ohkura, M., and Murakami, T. (1988). Direction judgement of blind travellers in straight walking under controlled environment. *Research Bulletin, National Rehabilitation Center for Persons with Disabilities*, Vol. 9, pp. 111-114.

(46) Nippon Foundation (1998). *Moudouken ni kansuru chousa* (Research about guide dogs).

(47) Ohkura, M. (1989). Studies on the mental workload of blind travellers by means of the secondary task technique. *The Japanese Journal of Ergonomics*, Vol. 25 (4), pp. 233-241.

(48) Tanaka, I, Murakami, T, and Shmizu, O. (1981). Heart rate as an objective measure of stress in mobility. *Journal of Visual Impairment and Blindness*, Vol. 75, pp. 55-63.

(49) Ikeda, H., Mihoshi, A., Ikeda, H., Ishizuka, Y., Horii, T., Nakaseko, M. (2006). Effects of barrier-free zone on walking and heart rates in before and after alternative of side-walk. *Traffic Science*, Vol. 37 (2), pp. 71-76.

(50) Iwahashi, H. (1983). *Shiranami ni muite: Miyake Seiichi wo kataru* (Toward white waves: About Seiichi Miyake). Traffic Safety Research Center.

(51) Ogami, H., Saito, S., Okura, M., Muraki, S. (2005). Investigation of physical S train and danger in sidewalk in stroller users: Focusing on cross and compound slope and level difference. *Journal of Life Support Engineering*, Vol. 17, p. 108.

(52) National Police Agency (2003). *Shikakushougaishayou fukasouchi ni kansuru secchi/unyoushishin wo seitei* (Establishment of setting and operative guidance regarding add-on equipment for visually impaired people).

(53) Mihoshi, A., Ikeda, H., Yanagihara, T., and Tabata, Y. (2004). Research regarding the introduction of LRT as a selective means of transportation: Application view points of the conversion from automobile/motorbike. *Journal of Asian Electric Vehicles*, Vol. 2 (2), pp. 633-638.

(54) Mihoshi, A., Ikeda, H., Yanagihara, T., Masui, A., Chikamori, M., and Tabata, Y. (2007). Confronting improvements of LRT in Japan. *Journal of Asian Electric Vehicles*, Vol. 5 (1), pp. 981-985.

Index

–A–
acoustic device 146
Activities of Daily Living (ADL) 18-19, 30
airplane 159
American Automobile Association (AAA) 68
Americans with Disability Act (ADA) 16
amputee 35
amyotrophic lateral sclerosis 33
armrest 50
arm 50
artificial ear 103-104
artificial parts of the body 35
asphalt pavement 62
assist device 75, 82-83, 86, 75-79, 161
Association of Technical Aids (ATA) 110

–B–
backrest 49
barrier-free 1, 13, 39, 59, 90, 153
Barrier-free Transportation Act 2, 13-14, 60, 67, 117, 139, 150
barriers on the road 14
block pavement 63
blood pressure 90
body belt 80
Braille block 131, 139-141
brain 109
brake 52
bus 16, 153-155

–C–
car 67, 72, 117, 120
care manager 19
caster 48
centrifugal force 88, 91-92
cerebral paralysis 33-34
clinical psychologist (CP) 21
communication 113-115
concave-convex surface 58
curb-stone 91
curve 93-95

–D–
decibel (dB) 101
defective colour vision 127
denial period 23
design 43-44

desire for affiliation and affection 9
desire for security and stability 9
disability 25
disuse syndrome 36-37
drain ditch 63
driver 67, 69
driver's license 6-71, 118-122
driving 87, 117-118, 123-124
driving school 69
duchenne muscular dystrophy 32

–E–
ear-trumpet 111
economic independence 9
educational history 113
elderly 1, 128-129
electric wheelchair 42, 44, 45, 54-55
elevator 151-152
escalator 152
evaluation 59
eyesight 129

–F–
facioscapulohumeral muscular dystrophy 33
financial aid 95-96
footrest 50
footstep 64
frame 53
fundamental plans for the facilitation of transportation 13

–G–
Gold Plan 11
gradient 60, 65
grating cover 63
grip 47, 52
guide dog 135-136, 138, 145-146
guideline 15, 60

–H–
hand rim 47, 65
hardware 2
healthcare cost 1
hearing aid 103, 104-112, 117, 123
hearing disability 13, 100, 103, 113, 115, 117, 122
hearing impairment 97
hearing loss 98-100, 108
Heart Building Law 27, 117
heart rate 59, 136-137
hemiplegia 8
hierarchy of needs 11

–I–
impaired autonomic nerves 31
impaired excretory function 31
impaired eyesight 126
impaired motor function 29
impaired sensory function 31
impaired trunk function 31
Independent Living (IL) 6, 20
information 115-116, 123
injury 54
Instrumental Activity of Daily Living (IADL) 19
interlocking block 62
internal impediments 1
International Year of Disabled Persons 6

–J–
Japanese Basic Act for Disabled Persons 7, 25
Japanese Industrial Standards (JIS) 42, 54, 56-57
Japan External Trade Organization (JETRO)

44
joystick 44, 54, 79

–L–
Law Concerning Assistance Dogs for the Disabled 145
license system 121
lift-up passenger-seat 86
light-rail transit (LRT) 156-158
limb girdle muscular dystrophy 32
limb/trunk dysfunction 1, 26-28
lip-reading 114
Long-term Care Insurance Act 11
lumber disk hernia 12

–M–
manual wheelchair 43-47, 54
medical social worker (MSW) 21
mental 1, 9, 12
module-model 47
motor function 27
motor paralysis 27
muscle activity 92-95

–N–
nerve 26, 28
New Barrier-free Act 14-15
normalisation 6-7
nursing care 10-13

–O–
occupational therapist (OT) 21
Order for Enforcement of the Building Standards Act 65
order-made model 46

–P–
paralysis 27
paraplegia 27
park and ride (P&R) 158
parking 72-73, 158
PDCA cycle 2
period of acceptance 24
period of confusion 23
period of effort towards a resolution 23
personal independence 9
persons with a hearing disability 97
persons with disabilities 5
persons with limb/trunk dysfunctions 26, 45
phantom limb 35
physical disability 1, 25, 101, 126
Physically Disabled Persons Welfare Act 25
physical problem 90
physical therapist (PT) 21, 46
physiological needs 9
platform 149
posture 66, 87-89
product configuration 43
progressive disorders 32
progressive muscular dystrophy 32
puddle of water 62

–Q–
quadriplegia 27
Quality of Life (QOL) 19-20

–R–
ramp 65
ready-made model 45
regular job 26
rehabilitation 20-22
remodelling 95
road condition 123
road surface 61

Road Traffic Law 55, 67, 118, 120, 121, 144
Road Transportation Act 155

–S–
seat 44, 49, 79-81, 95
self-actualisation 8
self-reliance 6
self-respect and desire for respect 9
senior car 55
sensory paralysis 27
Services and Supports for Persons with Disabilities Act 17-18
shock period 23
sidewalk 58-60, 61-64, 136-137, 141-143, 161
sign language and manual alphabet 113
skin 36
slope 64-66
social welfare 9
software 3
sound 109
specialist job 10
special mark 71, 120
special transportation service (STS) 155
speech disability 1
speech therapist (ST) 21
spinal cord 28-32, 36
spinocerebellar degeneration 33
station 16, 64, 149-151, 158
steering wheel 81, 94
subsidy program 74
super-elevation 85
support tool 137

–T–
Ten-year Strategy for Promoting Geriatric Health and Welfare 11
ticket-gate 149
tipping lever 53
traffic-light machine 146-147
train 149-151
transcript of a summary 84
transferring 84-85
turning device 81

–U–
uneven surfaces 60
universal design 14, 16-17

–V–
vehicle 158
visual disabilities 128
visual field defect 126
visual impairment 1, 60, 125-128, 130-132
volunteer work 3

–W–
welfare program 5
welfare service 18
wheelchair 39, 44-46, 142, 149, 159
wheelchair lift 85
wheel 47
white cane 134, 137, 143-144
workload 11
World Health Organisation (WHO) 19, 102
writing 114